The Future of War

Blackwell Manifestos

In this new series major critics make timely interventions to address important concepts and subjects, including topics as diverse as, for example: Culture, Race, Religion, History, Society, Geography, Literature, Literary Theory, Shakespeare, Cinema, and Modernism. Written accessibly and with verve and spirit, these books follow no uniform prescription but set out to engage and challenge the broadest range of readers, from undergraduates to postgraduates, university teachers and general readers – all those, in short, interested in ongoing debates and controversies in the humanities and social sciences.

Already Published

The Idea of Culture Terry Eagleton
The Future of Christianity Alister E. McGrath
Reading After Theory Valentine Cunningham
21st-Century Modernism Marjorie Perloff
The Future of Theory Jean-Michel Rabaté
True Religion Graham Ward
Inventing Popular Culture John Storey
Myths for the Masses Hanno Hardt
The Rhetoric of RHETORIC Wayne C. Booth
The Future of War Christopher Coker
When Faiths Collide Martin Marty

Forthcoming

What Cinema Is! Dudley Andrew
The Battle for American Culture Michael Cowan
The Idea of Evil Peter K. Dews
Television: Literate at Last John Hartley
The Idea of Economy Deirdre McCloskey
The Future of Society William Outhwaite
The Idea of English Ethnicity Robert Young

The Future of War

The Re-Enchantment of War in the Twenty-First Century

Christopher Coker

BLACKWELL PUBLISHING
350 Main Street, Malden, MA 02148-5020, USA
108 Cowley Road, Oxford OX4 1JF, UK
550 Swanston Street, Carlton, Victoria 3053, Australia

First published 2004 by Blackwell Publishing Ltd

Library of Congress Cataloging-in-Publication Data

Coker, Christopher.
The future of war: the re-enchantment of war in the twenty-first century / Christopher Coker.
p. cm. — (Blackwell manifestos)
Includes bibliographical references and index.
ISBN 1-4051-2042-8 (alk. paper) — ISBN 1-4051-2043-6 (alk. paper)
1. War. 2. War—Forecasting. 3. World politics—21st century.
I. Title. II. Series.

U21.2.C6397 2004
355.02—dc22 2004004239

A catalogue record for this title is available from the British Library.

Set in 11.5/13.5pt Bembo
by Graphicraft Limited, Hong Kong
Printed and bound in the United Kingdom
by MPG Books Ltd, Bodmin, Cornwall

The publisher's policy is to use permanent paper from mills that operate a sustainable forestry policy, and which has been manufactured from pulp processed using acid-free and elementary chlorine-free practices. Furthermore, the publisher ensures that the text paper and cover board used have met acceptable environmental accreditation standards.

For further information on
Blackwell Publishing, visit our website:
www.blackwellpublishing.com

Contents

Foreword

When Mark Twain wrote his satire, *A Connecticut Yankee in King Arthur's Court*, in 1889, he envisaged what the new tools of industrialization would wreak upon a western military culture steeped in the romance of war as a chivalrous calling. Twain's Colt revolver superintendent, Hank Morgan, magically transported to a medieval Arthurian world, uses industrial technology to destroy a feudal civilization, annihilating lines of charging armoured warriors with firepower and electricity. Only a magic spell from the necromancer Merlin that condemns the Yankee to a sleep of the centuries in a cave of the dead, finally ends the slaughter.

In the two World Wars, Twain's Yankee appeared symbolically on the fields of Europe to cut down the armies of the Old World with machine gun bullet, shell and aerial bomb. In these struggles, the industrialization of the twentieth century ended the mystique of armed conflict as a heroic and romantic endeavour in the spirit of Homer and Sir Thomas Malory. Mass production and battlefields of slaughter from Verdun to the Basra Highway disenchanted war in the twentieth-century human imagination. After such disenchantment, is it possible that the lure of war could yet again appeal under new conditions in the twenty-first century? It is the possibility of war's renewed appeal, its 're-enchantment', that is the subject of Christopher Coker's timely and insightful book on the military future.

In an era when so much of the speculative literature on the military future is technological and characterized by arid prose and a blizzard of acronyms, Professor Coker seeks to expand our cultural understanding of war. A distinguishing feature of this important study is not only its display of deep learning in philosophy and literature, but also its interpretation of the relationship between the instrumental, the existential and the metaphysical dimensions of modern warfare. These three dimensions seldom receive balanced attention in the literature of war, yet, as the author demonstrates, an understanding of their interaction is arguably the key to a sophisticated appreciation of the dynamics of war.

Developing his analysis on the connections between the instrumental, the existential and the metaphysical components of war, Professor Coker argues that, in the post-industrial twenty-first century, it is possible that war may have been re-enchanted by the power of cybernetics and biotechnology. Such new technologies allow advanced western states to contemplate a future in which the interface between human and machine permits the reduction of military risk, elevates the use of remote control networks over masses of troops and facilitates the optimization, rather than the maximization, of destruction. Twain's Hank Morgan, if transported forward into the twenty-first century, would find a precision and discrimination available in the tools of war beyond his wildest imagination.

Professor Coker believes that the future of war lies in biotechnology. He poses a key question: can war be re-enchanted by the manipulation of the biological over the cultural? Marshal Maurice de Saxe's famous saying that everything about war begins and ends in the human heart was meant as a cultural caveat, but in the future, the manipulation of genetics may permit changes in behaviour that transcend culture. In coming decades, our waging of war is likely to be facilitated by the means to produce bio-enhanced soldiers, by pharmacological abilities to manipulate combat endurance and by a computerized view of war that, in its mediated reality, will create disassociation by distance. These factors and the growing interface between human and machine, the rise of cybernetics and robotics, will pose challenges to the existential and metaphysical dimensions of war. New technologies simultaneously permit a humanizing of war by discriminate targeting and yet a dehumanization by disassociation.

The great challenge will be to retain the asceticism of the military profession, to understand that the human dimension of war that is expressed in existentialism and that metaphysics continue to confer meaning on action. As this book makes clear, sacrifice, the mark of the warrior, is a social phenomenon that is performed in defence of values. There could be nothing more stark in today's world than the contrast between the instrumental post-modern West and its struggle with existentialist Islamo-fascism and its cadres of suicide terrorists who seek meaning in life by death. The present struggle on the ground in Iraq between western soldiers and Iraqi insurgents and Islamic terrorists is a reminder that war is, ultimately, a profoundly human activity.

To win the wars of the future, it is almost certain that the West will have to rediscover the existential values of the military profession and metaphysical dimension of war. Twain's Yankee, proficient only in the technological tools of his age, will not be enough. In this respect, Professor Coker's identification of a need for a historical anthropology of ideas in order to allow us to locate our position in the *Zeitgeist* is a powerful reminder that wars occur because of human beliefs. For all those who seek a deeper understanding of the contours of war than that which one finds in standard journals of military strategy and technology, this fine book should be required reading.

Dr Michael Evans
Head, Australian Army Land Warfare Studies Centre,
Royal Military College, Duntroon

May 2004

Preface

> The formation flew backwards over a German city that was in flames. The bombers opened their bomb bay doors, exerting the miraculous magnetism which shrunk the fires, gathered them into cylindrical steel containers, and lifted the containers into the bellies of the planes. The containers were stored neatly in racks. The Germans below had miraculous devices of their own, which were long steel tubes. They used them to suck more fragments from the crewmen and planes. . . .
>
> . . . When the bombers got back to their base, the steel cylinders were taken from the racks and shipped back to the United States of America, where factories were operating night and day, dismantling the cylinders, separating the dangerous contents into minerals. Touchingly, it was mainly women who did this work. . . .
>
> The American fliers turned in their uniforms, became high school kids. And Hitler turned into a baby, Billy Pilgrim supposed. That wasn't in the movie. Billy was extrapolating. Everybody turned into a baby, and all humanity, without exception, conspired biologically to produce two perfect people named Adam and Eve, he supposed.[1]

This passage comes from Kurt Vonnegut's *Slaughterhouse-Five*, one of the seminal novels to come out of the experience of World War II. It describes an experience the author knew at first hand: the strategic bombing of German cities which began in earnest in 1941. Vonnegut was an 18-year-old Army Private when he found himself a prisoner of war in Dresden in March 1945 when the Allies conducted one of the last and most devastating air raids of all. At least 35,000 civilians were killed, most of them asphyxiated in basement bomb shelters. Unfortunately, prisoners like Vonnegut who had the grim duty of burning the bodies did so without the benefit of the daily ration of brandy issued to German troops.

Vonnegut's tale is partly science fiction, and in his tale history partly goes into reverse – in this case offering us the promise of the end of war. For centuries people have dreamed of closing Pandora's box in the hope of returning to a prelapsarian past. Is that day dawning? Are we evolving out of the war business? Was its late modern disenchantment a trend? Is the end of war on the horizon?

Those who believe so are likely to be disappointed. Let's go back to the bombing of Dresden. A press release issued by IBM in January 1998 described a software package which allows a three-dimensional visualization of objects on a screen. With it the company is helping to restore Dresden's Frauenkirche, one of the numerous buildings destroyed by the fire bombing of the city in 1945. Its software will allow as much as possible of the original – some 3 per cent – to be put back in its original place during reconstruction. It is a programme that promises a much better world, one in which not even the destruction of war is final. Unfortunately, another IBM press release on the same day reports another innovation using the same visualization software, but this time for a very different application: to help trim production costs, shorten throughput times and optimize process change. And the programme? The guided missiles of the future.[2]

Before we get too depressed it is worth recognizing that if war itself still has a future for the western world – the sole focus of this study – this is due largely to technology, especially the new technologies associated with the information revolution. It is that revolution which now offers the West the chance to reinvent war and fight it more imaginatively (and yes, even humanely) than in the past.

Vonnegut tells us that the experience of that terrible night in March 1945 when Dresden was torched in the closing months of the war was so traumatic that it took him another 20 years to recount the event. In the popular imagination, the dubious morality of strategic bombing in World War II was merely the most graphic manifestation of a much larger phenomenon that dated from the 1870s – the industrialization of war. On the industrialized battlefields of the Western (World War I) and Eastern (World War II) Fronts, war appeared to have become suicidal, or at the very least self-negating, an activity from which neither statesmen nor soldiers could derive status or profit.

War lost its appeal in the industrial age because it was seen as irredeemably dehumanizing. By the end of World War I, one of the last of the heroic, popular warriors – T. E. Lawrence – was able to talk of 'the distilled bitterness of a generation shot to pieces by war'. In an age of maximum firepower soldiers were reduced to human *matériel* (to use the contemporary jargon). The mechanistic principles of the day reduced the existential dimension of war to a minimum. The soldier increasingly saw himself as a cog in a giant wheel, an extension of the machine. And in terms of sacrifice (the meaning accorded to death on the battlefield), the great political ideologies of the late modern era demanded martyrs, not only by the thousand but by the million.

The bitterness Lawrence expressed continued to fester in the years that followed. And indeed, it still largely prevails. For some years now critics have been claiming that war as an *ontological* phenomenon is coming to an end. Theories abound as to why this is the case: ranging from the institutional (those who argue that war is being outlawed, or delegitimized) to the cultural (those who contend that it is likely to fade away, like duelling and slavery, two other social practices once thought to be central to the human condition).

Even warriors, as we traditionally understood or even admired them, would appear to be disappearing as a distinctive breed of men. For many western soldiers war has become a procedure, more cerebral than visceral, and almost entirely instrumental. Today's fighter pilots too, we are told, have become technicians for whom combat missions are now stripped of passion and danger. The central thesis of this book is that all such theories – whatever their empirical merits, and so far I believe they are not proven – are themselves a reflection of a fairly recent phenomenon (post-1870): the *disenchantment* of war.

The phrase comes from the pen of Max Weber, who wrote that the world had become disenchanted at the end of the European Enlightenment because life – robbed of mystery and magic – had been stripped down to a matter of rational calculation. This too was true of war, which through relentless advances in technology (especially firepower) was increasingly devalued even in the eyes of professional soldiers, let alone the unwilling conscripts called up by the million. For all of them, professional and amateur alike, modern warfare seemed to have become the supreme drama of a completely mechanized society, like that guyed so mercilessly in the 1920s film *Metropolis* with its ominous scenes of city skyscrapers intercut with factory wheels and the face of a huge clock evoking the discipline of time keeping.

What I will contend in the following pages is that war has been *re-enchanted*. And the principal reason has much to do with the fact that we live in a post-industrial age. Of the three revolutions that have dominated our imagination, the atomic may well have led war into an end game. The information and biotechnological, however, would seem to have invested it with a renewed lease of life, if not a determination to play the game a little longer by other rules.

Digital biology is likely to be the key to the future in every walk of life. For the two revolutions are not distinct. Indeed, the decoding of the human genome would have been impossible without the increase of computing power provided by the information technology (IT) revolution. Genetic manipulation required the decoding and recombining of the information codes of living matter which was only made possible by an exponential increase in processing power. Conversely, the language of the IT age has been significantly influenced by nature. What the genome project revealed is that we have almost the same number of genes as the chimpanzee. What makes us different – what makes us the intelligent creatures we are – is the networking and recombining capacity of our cells, and particularly our brain cells, through millions of electrochemical connections. It would seem that in terms of networking and feedback loops (the basis of cybernetics), the human brain is similar to the Internet in our computer-run societies. In an attempt to create more complex computer brains, scientists are also studying complex neural networks in the human brain in the expectation of constructing 'digital chromosomes' with many of the same features as our own DNA.

In time scientists who understand the processes of nature (especially those who know how complex adaptive systems work) will be able to build computers that can evolve (not solve) most conceivable problems. In computer programming we talk of 'evolutionary algorithms' (programs that permit things to evolve in computer space). We are already attempting to evolve digital 'immune systems' within computers to fend against hacking by cyber-criminals or terrorists.

When it comes to the warrior, not just war, digital biology is also redrawing the rules. *Existentially*, the warrior is now a subject that can be enhanced through cyborg technologies and genetic re-engineering. It is called the 'post-human' condition, a term that suggests that the traditional split not only between man and machine, but also between man and nature, is fast disappearing. Nature is being modified by technology, while technology, in turn, is becoming assimilated into nature both as a function and as a component of organic life. In challenging the fixity of human nature, the biotechnological age is forcing us to question many of our assumptions about personal identity, including what constitutes the grounds of our uniqueness as a species. Not only is this true for life in general, it is also becoming true for war.

What this book sets out to do is to illustrate this process. What it argues from the evidence presented is that in two crucial respects the information and biotech revolutions have changed the way we think about war; they have begun to change the way we practise it, too. Instrumentally, its humanity is to be found within the technological advances that the IT age has made possible, in the form of weapons systems which can be targeted more precisely than ever, in a way that promises less 'collateral damage' than ever before. But the potential of biotechnology also promises to transform the existential dimension of war by 're-engineering' rather than, as previously, merely retooling the soldier, so that his abilities (not only capabilities) are enhanced. For it would be quite wrong to imagine that genetic re-engineering, on a large or small scale, will make human beings 'other' than we are, or that in some quasi-Nietzschean sense it will enable us to 'overcome' our own humanity. In altering human beings, as they almost inevitably will, scientists will merely allow us to do what we have always done, but to do it better. And war is what some will continue to do very well.

1

The Re-Enchantment of War

Ruskin and the Disenchantment of War

> The odds is gone
> And there is nothing left remarkable
> Beneath the visiting moon . . .
> *Anthony and Cleopatra*, 4.15.66–8

Where else to start but with that most symbolic event of an age, a paean to progress and, above all, the idea of Progress, the Great Exhibition of 1851. There was much to admire for the millions who visited it over the next few years, especially the crowds that were ferried to London by the new railways, many of them from the industrial cities of the Midlands and the north. Particularly inspiring were the fifteen cotton spring machines as well as the steam turbines and locomotives, all of which were seen as gleaming incarnations of progressivism. These and other wonders were taken to be – as James Ward wrote at the time in *The World and its Workshops* – 'the epitome of man's industrial progress – of his untiring efforts to release himself from his material bondage'.[1]

Opening the Great Exhibition, Prince Albert too had looked forward to the 'realization of the unity of mankind'. Geographical distances which had previously separated the different communities and parts of the globe were rapidly vanishing before the achievements of modern technology, and so too, it was hoped, the cultural divisions which they bred as well. This was the peace dreamed of by utilitarians such as Jeremy Bentham and liberals such as John Stuart Mill who welcomed the fact that the marriage of science and capitalism promised future peace. Others, of course, also saw in that marriage the future evolution of war.

Indeed, there were a number of critics at the time who were remarkably unimpressed by the Great Exhibition. One was the Scottish historian Thomas Carlyle, another the art critic John Ruskin, soon to become a fiery prophet, his rich and multivalent prose levelled against the abuses of the industrial age. Both men were among the first to acknowledge that the revolution had begun to liberate humanity from the worst aspects of manual work, from disease and poverty. But while it lifted the suffering of the body, the industrial revolution, they believed, had also imposed a heavy burden on the *soul*. This was especially true of the human contact with God, nature, the earth and even human nature (psychoanalysis was an invention of the industrial age). Modernity may have eased the burden of being human: the burden of work and, for women, the burden of procreation; it may have helped abolish exacting manual labour, the worst diseases, and even the terror of old age, but it was also incredibly soulless.

Why war was soulless

'Men were not intended to work with the accuracy of tools', Ruskin was later to complain:

> to be precise and perfect in all their action. If you will have that precision out of them and make their fingers measure degrees like cogwheels and their arms strike curves like compasses, you must unhumanize them. All the energy of their spirits must be given to make cogs and compasses of themselves.[2]

'Engine turn precision' had by then, or so it appeared, become mankind's destiny – not the destiny of humanist subjectivity (of the man who is determined from within), but destiny from without (the logic of technology, or the determinism of science).

If Ruskin had little love for the industrial age, he had even less for the age of industrial warfare. Perhaps it counted for much that he himself was not untouched by war itself. Only a nation that produced great soldiers, he believed, could also produce great art. But great art, of course, could also fall victim to war. As he reminded his readers in the year that they flocked to the Great Exhibition to see the fruits of the industrial revolution at first hand:

> The greatest pictures of the Venetian masters were rotting in Venice in the rain for want of a roof to cover them, with holes made by (Austrian) cannon shots through their canvas.[3]

Ruskin had no personal experience of the battlefield but, ever the discriminating art critic, he was particularly saddened by the loss of three of his beloved Tintorettos in the Franco-Austrian War of 1859.

But what depressed him most was the knowledge that war had succumbed to the scientific dynamic of the age. Seeing science as he did as both agent and symptom of moral decline, he regretted the fact that industrialization had devalued war just as he believed it had devalued life. The mechanization of war, he wrote in 1866, had denied it a higher meaning. War had become a measure of which nation had the largest gun, or the best gunpowder made by the best chemists or the best iron smelted with the best coal. War had become a matter of 'machine contriving'.[4] His distaste for the mechanization of war was part of his protest at the industrial revolution itself. Just as 'engine turn precision' had destroyed the dignity of human labour, so machine contriving had destroyed the dignity of soldiering as well: war was no longer 'a pure trial of wills', a test of character, a duel between warriors, but a matter of impersonal calculation.

A few years later, at the time of the Franco-Prussian War, another contemporary writer, the novelist Georges Sand, was equally struck by how soulless war had become. Following the battle of Sedan, the war's culminating point, she complained:

> This war is particularly brutal, without soul, without discernment, without heart. It is an exchange of projectiles in greater or lesser number, of greater or lesser range, which paralyses worth, nullifies the soldier's awareness and will. No heroes any more, just bullets.[5]

Two critics of the times, the same judgement. Their critique was not at variance with Max Weber's analysis of the age, and what he described as 'the disenchantment of the world'. What he meant by the term was the increasing technological rationalization of modern life. The religious worldview which the Enlightenment philosophers had destroyed had been an enchanted world filled with magic and the mysteries of God. Cold, hard rationality had killed off both. Even for those who still possessed a faith, the world had ceased to be magical. Everything was now accountable to science, scientific axioms and mechanistic rules.

> The fate of our age with its characteristic rationalization and intellectualization and above all the disenchantment of the world is that the ultimate most sublime values have withdrawn from public life either into the transcendental realm of mystical life or into the brotherhood of immediate personal relationships between individuals.[6]

Modernity had brought to an end a sacred sense of wholeness, the reconciliation of self and the world provided by myth, magic, tradition, religion, the numinous and even the 'immanent' in nature. Wilhelm Hennis has called this 'the central question' in Weber's thought: the relationship between the ways in which we attempt to give meaning to our existence and the constraints imposed by the logic of power and modernity. Years later Freud too was to complain of the loss of that 'oceanic feeling' that he placed at the centre of his own book, *Civilization and its Discontents*. It was a sentiment found in many other works from Durkheim to Simmel. All of them recognized that even the reflection on what was perceived as lost or the effort to cope with the sense of loss had become synonymous with modern thought itself.[7]

In the specific case of war the discipline required of mass armies also demanded the rational conditioning of performance, or uniform and predictable behaviour. What Weber said of life in general Ruskin and others would have said of modern war, that in adjusting to the demands of the world, to its tools and machines, humanity had been reduced to an industrial function. Until the early nineteenth century, war had been enchanted not by gods or spirits but by heroes and heroic action, by the existential realm in which the best (as well as the worst) of human behaviour was exhibited. Now war had become a matter of rational calculation.

It was the fate of Ruskin and his contemporaries to witness the first modern conflict, the Crimean War (1854–6), which broke out only three years after the Great Exhibition. In some respects, of course, the war was a traditional one. The very first battle, Alma (1854), would have been familiar to the Duke of Wellington, who had died three years earlier, with soldiers in red tunics marching in file, led into battle by generals on horseback – following the generals, not being despatched by them into battle, as was to be the case in the future. Within three months, however, trench warfare had begun in earnest. The Crimean War was the first to see the appearance of a modern medical commissariat, the first to be covered by war reporters and the first to attract popular interest and initial enthusiasm.

And it also witnessed a level of inhumanity that was ominous for the future. Several weapons proposed but never deployed would go on to help define twentieth-century conflict. Plans for chemical warfare against the town of Kronstadt were discussed by the British government, as was the use of a kind of prototype tank, a locomotive land battery fitted with scythes to mow down infantry, though nothing came of either proposal. As Sebastopol was about to fall, one Russian officer regretted that 'the means of destruction of people are getting stronger with each day; and it would be no wonder if they invented some sort of machine which will kill a thousand people in one go'.[8]

War as a cultural phenomenon

To understand why war as a phenomenon had become disenchanting, we need to apply the categories of a phenomenologist. We need to understand what war is in order to see what it was rapidly becoming. In the language of Clausewitz (the first phenomenologist of war, and by far the most perceptive), war was in danger by the early twentieth century of contravening its true nature.

For whenever we speak of 'war' we are talking of a very complex human activity which is made up of three interrelated dimensions.

1. As an *instrumental* concept war refers to the ways in which force is applied by the state, the way in which it is used to impose one state's will upon another. War, as such, is a rational instrument employed by states in a controlled, rational manner for purposes that are either economic or political.
2. As an *existential* concept the term refers to those who practise it: warriors. As Hegel always insisted (contra Kant), war would only end when warriors no longer needed it to affirm their own humanity.
3. As a *metaphysical* concept war translates death into sacrifice – it invests death with a meaning. And it is the metaphysical dimension which is the most important of all precisely because it persuades societies of the need for sacrifice. It is sacrifice which makes war qualitatively different from every other act of violence. We rarely celebrate killing but we do celebrate dying when it has meaning, not only for the dead, but for those they leave behind.

Now, for war to have been disenchanted, each dimension should have been devalued in the course of the early twentieth century, and indeed each was in its own unique fashion.

Instrumental

War is the realization of human potential: it represents a supreme act of will. It ceases to be tribal for it is no longer about blood feuds, or waterhole rights; it is not determined by factors extraneous to ourselves such as food scarcity. Instead, war is an act to compel others to do our will. The state is the unit by which humanity wills its history. And until very recently war was the main instrument of the 'will to power'.

As Nietzsche recognized, the move from cultural to social selection – the transition from hunter-gatherer societies to the state – was 'the fundamental change' in human life, the caesura between natural selection and self-conscious evolution that 'occurred when (man) found himself finally enclosed within the walls of society'. For it was this move which transformed 'the semi-animals' we once were into the human beings we have become. Society, wrote Nietzsche, gave rise to 'an animal soul turned against itself' and thus to a new species 'pregnant with a future'.[9]

War, in other words, came into being with social not cultural selection. Primitive tribes may be in a constant state of war (state of nature) – each tribe maintains its identity by being at war with others. Should the opportunity arise, one group may seek to expel, or even eliminate, the other. But the state, by contrast, rarely eliminates its enemies; it absorbs them. Once it arrives on the scene its function is to expand or assimilate. Power requires the subordination of others. Which is why the emergence of the state is the main event in human history not only for the reasons favoured by sociologists – it shows a stage in the progressive differentiation of social functions and the stratification of status groups. It also corresponds, as Nietzsche tells us, to a massive revision in human *possibility*.[10]

In the course of the twentieth century, however, the state's very existence was imperilled by its attempt to realize the impossible: that one state, representing one branch of humanity, should permanently seize the high ground of history. Every nation considered itself to be the custodian of truth, history, or a national destiny that had to be tested on the battlefield – and the prophets were there to instil in them the spirit of sacrifice: Guizot in France, Fichte in Germany, Mazzini in Italy, and a host of lesser prophets who addressed their prophecies to the Slavs.[11] All such claims demanded that their legitimacy be tried in battle, and that those martyred to them should be numbered in the millions.

The upshot of all this was that wars became even more deadly and destructive. The quintessential battle, perhaps, was that of Verdun (1916). By the time it ended, every square metre of the battlefield had received about 1,000 shells. Not for nothing did the German commander call this 'the mill of the Meuse'; the human beings caught in it were ground exceedingly small. Verdun was an entirely new battle: one of annihilation. And the disenchantment it produced could not be any greater than that suggested by the name the Germans chose for the operation – *Gericht*, a word translated by Alistair Horne as 'execution ground'.[12] Of all the battles of the twentieth century, Verdun was probably the most appalling in terms of its human cost.

The Italian journalist Curzio Malaparte put the case against war very well in a conversation that appears halfway through his great novel *Kaputt*, a vivid portrait of Europe at the end of World War II as a continent that by then seemed intent on its own destruction.

> 'The greatest problem of modern days is still the religious problem', said Bengt von Toine. 'On ne peut pas tuer le Dieu – God cannot be killed.' . . .
>
> 'The modern state', said Constantinidu, 'deludes itself into thinking it can protect God simply with police measures.'
>
> 'It is not only God's life. The modern state deludes itself that it can protect its own existence', said de Foxa.[13]

The craving for war did not necessarily die in the ruins of Verdun – as Hitler's war showed, it remained alive in the hearts of those who ran the Third Reich, but with their defeat the state's invincible faith in war as the realization of human possibility could never be rekindled.

Indeed, Europe's retreat from war can be traced to the early 1930s when many of the French veterans who returned to the battlefield of Verdun chose to celebrate it, not so much as a French victory, or a German defeat, as a European catastrophe. It has been remembered as such ever since, and as such it has become one of the founding myths of the European Union, an institution which aspires to *transcend* the nation state in the name of a political culture quite new in history: a transnational community, a Kantian cosmopolitan society that has largely renounced war by living at peace with itself.

The existential warrior

War itself would not survive if it were just instrumental. The state has to rely on those who are willing not only to kill for a state-sanctioned cause, but to hazard their life in battle. The true warrior, wrote Hegel, affirms his humanity in a way that is specific to himself, for he fights to impose his own worth on the world. The stake is no longer a disputed prey (the origin of human conflict) so much as the opinion another holds of him (the origin of war). The warrior is one who prizes the recognition of his fellow men above survival and is willing to hazard his life (as well, of course, as the lives of others) in order to secure it. He fights for honour or a medal, or a flag – not because of their intrinsic value, but because they are desired by others.

In short, the true warrior is a human type and war will only end when that social type is no longer honoured by his fellow men. Warriors live in the recognition of their fellow citizens; in the story told of their lives after they are gone; in the esteem in which they are held by their bravest enemies (those whose esteem is worth having). Exclusive to the human species is desire – we desire not what we want but what others esteem. In that sense desire is not an appetite, it is not instinctual: it is social. In some cases we desire to outdo others – to rob them of what we wish to possess: their reputation. What we desire, above all else, is respect, and it is through the warrior's conduct in battle that respect is won.[14]

Not that we should exaggerate the warrior's indifference to death. No soldier – not even a warrior – wishes to die. What the soldier really celebrates is his survival. Even Achilles's frenzy to kill, Harold Bloom reminds us, is 'a dialectical process against mortality itself'.[15] What we find in Achilles is a zest for life, not a willingness to throw it away in battle. Even in Hades, when Odysseus visits him on his journey home, Achilles is still resentful of death, even the early death he chose for its fame, instead of the obscure but long life he might have enjoyed had he never joined the Trojan expedition. Yet there is no more powerful evocation of what makes the warrior different from others than Achilles's reminder that we all owe death a life, a life that will be remembered long after we are gone. Consult the adaptation of Homer's epic by the poet Christopher Logue and you will find this wonderful description: we hear the light of Achilles's helmet 'screaming across three thousand years'.[16]

But there is another kind of dialectics in war. 'Deep under the areas where the dialectics of war are meaningful', wrote Ernst Junger, that most emeritus of twentieth-century warriors, 'the German met with a superior force: he encountered himself.'[17] The same can be said of Junger's own writings, which offer a fascinating insight into the born warrior's mind. For Junger conforms to the warrior type described by Glenn Gray (a US counter-intelligence officer in World War II), who published the most enlightening study of warriors ever written. The delight in war as spectacle, the delight in comradeship, the delight in danger, are all to be found in Junger's life. As Gray puts it, battle can often be 'the one great lyric passage in a man's life'.[18] And it was Junger's gift as a lyricist which enabled him to write about the Great War in quasi-Homeric terms.

There is an entry in his World War II diary which runs as follows:

> War isn't like a cake that the two sides divide up between them to the last crumb; there is always a piece left. That's the piece for the

> gods and *it remains outside the argument*, and it elevates the fighting from sheer brutality and demonic violence. Homer knew it and respected it.[19]

This is the existential element outside the argument (the causes of the Trojan war), outside the instrumental reason why it is fought. Marx had asked: 'Is Achilles possible with gunpowder and lead?' Junger struggled to provide an answer in the affirmative; indeed, all his war writings are an attempt to affirm the existential side of war even in the midst of slaughter on a truly inhuman scale. His classic account of his own experience, *Storm of Steel*, was an attempt to capture the existential experience of war on the Western Front, which, like Troy, could be seen as a great fortress that was invested for four years, not ten. And there are even some stirring Homeric passages in the book – if we care to look for them:

> A figure in a brown waterproof came quietly across the fire-swept piece of ground and shook hands with me. Kius and Boje, Cpt Junker and Schaper, Schrader, Schlager, Heins, Findeisen, Hilemann and Hoopenrath stood behind a hedge that was being raked with lead and iron, and held a great council of war. On many a day of wrath we had fought on one battlefield together, and this time, too, the sun, already low in the west, was to gild the blood of all, or nearly all.
>
> It was our last storm. How often in years gone by we had stepped out into the western sun in a mood the same as now. Les Esparges, Guillemont, St Pierre Vaast, Langemarck, Passchendaele, Mouevres, Vracourt, Mory! Again the carnival of carnage beckoned.[20]

All his life Junger struggled to affirm the warrior virtues on the industrialized battlefields of the early twentieth century. Ultimately, he failed – or rather, his account does not ring true because there really wasn't anything Homeric about World War I. Or only in the subtext of Homer's story – in a reference, for example, to the 'funeral pyres burning night and day', so numerous are the casualties even in one day's fighting. The poem, in another memorable phrase of Christopher Logue's, is 'all day permanent red'.[21] For notwithstanding the graphic scenes of the death of heroes such as Hector and Achilles, Ajax and Paris, even in the *Iliad* the main death toll is clearly to be found in the ranks.

In her own take on Homer's poem Simone Weil (writing in 1940) found nothing but unremitting warfare so consistent with the unrelenting warfare of the century in which she lived. She identified force rather than heroism as the main protagonist of the epic – all the heroes, including Hector and Achilles, are transformed into soulless beings. For if force is the hero, it hardly matters who is wielding it – or why.[22] Even without the overly self-referential nature of Weil's work, her 'reading' of Homer is not unrepresentative of the modern experience of war. Despite Junger's own heroic deeds, and those of countless others like him, heroism could not make up for so many splintered minds and broken spirits.

Metaphysics

'There is only one thing I fear: not to be worthy of my suffering.' In that one sentence Dostoevsky encapsulated the third dimension of war. A true warrior must make sense of his suffering. His willingness to sacrifice his life makes sense of his life. If his death has meaning for others, then he can accept that death is his destiny – the natural completion of his life. At the moment of death, his life is transfigured.

Sacrifice sacralizes suffering. The word itself is derived from the Latin for 'sacred'. But going beyond its etymology, the Hebrew tradition suggests that suffering can be redemptive. It can be the foundation of a better world collectively, if not individually. And, of course, arising from that tradition, Christianity made martyrdom into a redemptive act. Clearly, the concept of sacrifice can be found in other religions, especially Islam, but it is nowhere as significant as it is in the western tradition that secularized it in the course of the nineteenth century.

For western philosophers taught that wars are won and lost in the end by a soldier's willingness to die, not only for the state, but also for what Hegel called an 'ethical idea': the idea which the state embodies. The sacred is all important because it is functionally necessary to many aspects of life, including war. De Tocqueville talked of the loss of the sacred as an 'aberration of the intellect and a sort of violation (of human nature)'.[23] The gods may come and go, writes Roger Scruton, but they live through our need for them and we make room for them through sacrifice. The gods come about because we idealize our passion, and we do this by sacrificing ourselves to the vision on which they depend. It is by accepting the need for sacrifice that we live a 'sacred' life and find meaning in our lives. Scruton puts it well when he writes: 'seeing things that way, we recognize that we are not condemned by mortality but consecrated by it'.[24]

Unfortunately, in the late modern age (1870–1945), nationalism proved to be the greatest devouring god of all. Bastille day, Armistice day, ceremonies for the fallen in battle – when it came to the ritual and ceremony of collective identification, nothing was more powerful than war. The nation state was a political religion and it was not fortuitous that the rise of nationalism should have coincided with the decline of Christian belief. As Hans Kohn writes, 'messianic dreams with the nation at their centre put the nation into immediate and independent relation with the Absolute, and one of the main ways in which religion and nationalism are structurally similar is their ability to sanction sacrifice for the Absolute'.[25]

The two most important messianic dreams – Marxism and fascism – gave war particular force. Hitler designated the ideological content of national socialism 'a philosophy of life', but, as many writers have observed, fascism, like Marxism, was less a philosophy and more a 'political religion'. In the case of Nazism, all manifestations of the social structure were interpreted as epiphenomena, the expression of a more primordial 'national' base, race or blood. Like all religions, Nazism derived its force from the claim that the irrational (spirit) would always prevail over the material (technology). Its message was that through will alone a nation could bridge the gap between the transcendental and the mundane; that though war was immanent in nature, a people could escape the inexorable workings of natural law (they could defy the material circumstances of life) through willing their own fate.

Even the democracies adhered to their own metaphysical myths. Although Junger criticized what he called the 'typically shallow French attitude' in anti-war novels such as Barbusse's *Under Fire* for stressing the purely material aspects of war, and not 'the responsibilities that demand sacrifice', Barbusse himself was thoroughly a hostage to a metaphysical idea: progress. 'I see too deep and too much,' Barbusse has his hero claim in his great novel, and what he himself saw in World War I, its butchery notwithstanding, was 'the lofty promise of final progress which leads true men to give their blood'.[26]

Unfortunately, the scale of sacrifice was too great in the twentieth century. Without the disclosure of an agent, soldiers lost their sense of agency, and hence found it difficult to inscribe their experiences with meaning. Unlike works of art which retain their relevance, whether or not we know the artist's name, action without a name, a 'who' attached to it, is largely meaningless. 'The monuments to the Unknown Soldier after World War I', wrote Hannah Arendt in *The Human Condition*:

> bear testimony to the then still existing need for glorification, for finding a 'who', an identifiable somebody whom four years of mass slaughter should have revealed. The frustration of this wish and the unwillingness to resign oneself to the brutal fact that the agent of the war was actually nobody, inspired the erection of the monuments to 'the Unknown Soldier', to all those whom the war had failed to make known and had robbed thereby, not of their achievement, but of their human dignity.[27]

For Arendt, it was not the World War I poets who came nearest to expressing the disenchantment of modern war so much as William

Faulkner's parable *A Fable* (1954), which aptly took as its hero, not a man with a name, but one without – the Unknown Soldier.

Disenchantment and the Industrialization of War

The increasing disenchantment of war as well as the struggle to keep the enchantment of war alive must both be seen in terms of the industrial revolution. And the origins of the industrialization of war can be traced back to the last years of the Napoleonic Wars.

For sheer waste of life the wars had little equal. Using a sampling technique and concentrating on selected regiments, one French historian has reckoned that the French alone lost 916,000 men – a figure that includes soldiers who died of disease, as well as those who went missing or died in battle.[28] The carnage for the other armies, though not as great, was still unprecedented.

The London *Observer* for 18 November 1822 reported that the previous year a million bushels of human and horse bones had been imported from the neighbourhood of Leipzig, Austerlitz, Waterloo and other battlefields. 'The bones of the hero and the horse which he rode' had been shipped to the port of Hull where they had been forwarded to Yorkshire bone grinders who had built steam engines for the sole purpose of granulating the bones and selling them off for fertilizer.[29] Here was a terrible intimation of the future industrialization of war. And it was appreciated not by Clausewitz but by his contemporary, Antoine Jomini, who had been acutely aware of the actual and potential importance of modern weapons technology. In a Sibylline passage he predicted that the weaponry of the future was likely to have a decisive impact on the outcome of war:

> The means of destruction are approaching perfection with frightful rapidity. The Congreve rockets . . . the Perkins steam-guns which vomit forth as many balls as a battalion . . . will multiply the chances of destruction, as though the hecatombs of . . . Leipzig and Waterloo were not sufficient to decimate the European races.[30]

What Jomini perceived as the quickening pace of technological evolution, had he but known it, was the application of the industrial revolution to war.

It is true, of course, that the Napoleonic Wars were not dominated by technological change, but they were the first conflicts in which the initial stirrings of technological inventiveness could be seen. The problem for the technologists themselves was that science still could not render their inventions practical. Robert Fulton's book *Torpedo Warfare and Submarine Explosions*, which was published in 1810, looked forward to the turn of the twentieth century – but it was to take a different scientific age to make submarine warfare possible. The French used manned balloons for observation purposes but their operational range was severely limited. Cavalry continued to remain the most valuable means of reconnaissance. As Shelley wrote at the time, 'the art of navigating the air is in its first and most helpless infancy. The aerial mariner still swims on bladders and has not mounted even the rude raft'.[31]

There were designs for steam-powered paddleboats to attack French shipping in port. Congreve's incendiary rockets were used against Copenhagen with devastating effect in 1807 when they killed several thousand citizens. But the age of rockets lay in the future, and the next city to be bombarded with rockets was London in 1944. Steam power was used in the manufacture of guns – a steam-powered plant in Birmingham manufactured 10,000 muskets and barrels a month, but, for the most part, the age of mass weapons production was yet to come.

It was the combination of the technological dynamic and Napoleon's use of cannon that created another disenchanting aspect of warfare: firepower. It was significant, perhaps, that Napoleon began his career as a captain of artillery for he himself put great emphasis on that arm. In claiming that 'missile weapons have now become the principal ones', he echoed Turpin de Crisse's statement that 'it's by fire, not by shock, that battles today are decided . . . (and) . . . it is with artillery that war is made'.[32] Battering at the enemy's line often for hours, the sheer quantity of fire concentrated was unprecedented for this era of warfare. A battery of only 40 guns, occupying a front of about 600 yards, might throw a thousand rounds an hour into an enemy position and still increase their rate of fire minutes before an attack.

Even so, the damage inflicted was not always as great as men recalled. A French gunner walking over the field of Wagram (1809) the day after the battle was disappointed to find how few lives had been taken proportionate to the ammunition expended, even though he thought the noise of the French cannonade the most deafening he had ever experienced. It was more demoralizing than dangerous for soldiers had no means of returning fire.[33] Confronted by death, they often had no prospect of coming face to face with their enemy, or engaging him hand to hand.

Yet it was not long before these deficiencies were remedied. The military-technical revolution of the nineteenth century began when western armies dropped the smooth-ball musket (essentially the same weapon in use since the late seventeenth century) and adapted the rifle, a breech-loading (as opposed to muzzle-loading) weapon which allowed infantry not only to load and fire more rapidly, but to do so without standing up. Rifled, fast shooting, infantry weapons were only the beginning of the trend. By the end of the century machine guns could deliver a higher rate of firepower than ever. By automatically completing in a single cycle all the operations previously executed separately – releasing the firing pin, opening the bolt, ejecting the cartridge, loading the new round and locking it into place – a machine gun could fire between 550 and 700 rounds per minute. Military skill was made redundant. The machine gunner became a true worker, the product of an industrial age.

Mass firepower became necessary because, transported by rail, millions of soldiers could also be put into the field for the first time. An entire society could be mobilized for war within weeks. 'I *was* going in the middle of September . . . to Ammergau to see the Miracle play', complained an English visitor in 1870, 'but the chief person has taken off to serve in the artillery with Judas Iscariot as his superior officer.'[34] A terrible cliché arose by the end of the nineteenth century, the 'war machine', which summed up, as no other term did, the transformation of an age-old custom by the industrial revolution. In time the mechanization of all aspects of life, which was greatly accelerated by war, left its imprint upon everyday language. The dehumanizing phrase 'human *matériel*', which was still denounced before World War I as a denial of the human spirit, soon became an accepted part of speech.

Ruskin's image of war as a duel between states was absorbed into a vaster image of a gigantic working process. Not only did war serve industry, it had itself become a vast industry. Recognizing this fact, Churchill described the divisions sucked into the battles of Verdun and the Somme as 'the teeth of interlocking cog wheels grinding each other'.[35]

Technology in the Modern World

All this is a good illustration of Martin Heidegger's claim that 'the essence of technology is not technological'. Its essence lies not in the tool or later the machine, but in the man who uses it. It resides in the way we think or imagine our external world and our relationship with it. Technology is not value free. How could it be? It is one way by which we project our own power – the bow extends our range, the computer amplifies our knowledge. Both enhance our power to will particular outcomes.

Now, what makes us unique as a species is not that we use tools, but that we attempt to remake the world, to mould the external environment to our needs. Indeed, except for the time we are asleep, we are always producing things to enhance our relationship with the external world. Heidegger memorably defined the life-world of man in terms not only of tools or machines but also of their use and users. For as the French ethnologist Marcel Maus also argues, we are the products of such use. As human beings we are always workers (our humanity, Hannah Arendt wrote, is defined by our work). We are technological beings engaged in technological activity.[36]

Inventiveness may not be unique to our species but even where other animals use rudimentary tools (such as chimps who use stones to crack nuts), they are largely the product of knowledge rather than imaginative reason. One learns inductively by trial and error or by mimicking the behaviour of older members of a colony. Without deductive reasoning, of course, there is no inventiveness.[37]

What makes technology innately inventive is a cognitive process – a way of seeing how to use or apply it that requires a mentality, or 'worldview'. The hand axe is a good example. As a tool it goes back at least 1 million years, but what is important about it is its standardized nature. What impresses archaeologists is that despite differences in regional 'style' and the different raw materials used, the hand axe is a fairly uniform instrument. It is a product of communicated knowledge, and to be communicated knowledge requires to be abstract. To reproduce a standardized piece of equipment over time, its makers must have some idea of a tool in general.

Even a crude toolmaker has to learn his skill and learning is artificial. Tool making is not a social mode of adaptation such as the nut-cracking propensities of a chimpanzee colony. Craftsmanship is inherently artificial in that it is not concerned with the necessary but with the contingent. It is not concerned with 'how things are but how things might be – in short, with design'.[38] In that respect, the 'becoming' of man depends inherently on tools or design. 'Man creates and at the same time he creates himself', writes Marcel Maus.[39]

Heidegger went even further in arguing that it is a man's relation to things (and not to other people) that defines his humanity. 'What is a thing is the question who is man?' And the reason for that is that technique (*techne*) brings forth man and unveils him to himself. *Techne* to the Greeks designated not only the making of useful objects, but also the 'bringing forth of truth into the splendour of radiant appearing'.[40] It was here that Heidegger drew a contrast between the technological dynamic of the pre-modern and modern worlds. In the latter, it is a making happen or a manufacturing of reality. In the pre-modern it was the bringing forth of what nature concealed. *Techne* for the Greeks (Heidegger invoked 'the Greeks' as a shorthand for the ancients) is the human unravelling of what is already there. The craftsman who designs a tool does so from what nature provides. No violence is done to nature itself. For nature is cosmic. It is part of the divine plan. To do violence to nature is to violate the holy as well as one's own humanity. In this view of technology the gods allowed man to reveal what was already in the natural world. The world, in short, achieved its final (god-given) design through human agency. As craftsmen, the Greeks can be said to have conceived of themselves as assistant workers cooperating with an artist – God – in the production of the ultimate artwork – the world.[41]

Let us take the example of Bronze Age craftsmen forging a shield, similar to the one forged by Hephaestus for Achilles. The craftsman forged the shield from material that was already there with the addition of human ingenuity: tin ore which, when added to copper, produces bronze. Human beings also invented the goatskin bellows which increased the temperature of the fire and thus enabled the forging of a greater quantity of metal ore, or shields en masse. And the craftsman designed the shield to a specific order, a divine design in furtherance of a human practice: war. For as we are told in the *Iliad*, 'so the blessed gods rallied the opposing forces, forced them together and opened up strong strife among themselves'. War between men, as Homer tells us, was part of the gods' design for the world.

The example illustrates two ways in which technology was part of an enchanted, not disenchanted, world. The ancient craftsman showed delight in fantasy and symbol. The shield of Achilles is replete with imagery that constitutes a microcosm of Greek life, its binary purpose – war and peace. The tendency towards economy and efficiency, towards functionalism, was counterbalanced by the desire to exhibit human value and purpose. In the modern era, by comparison, it is the function of an object which is all important, and that function refers back to utility, not to a larger spiritual world.

Secondly, the pre-modern age was one in which the craftsman worked with what nature had yielded to man. In the end, nature could yield her secrets but one could never prize them from her by force. One had to respect the rules deduced from observing her rhythms. One had to behave 'according to nature,' never 'contrary to her'. In the pre-modern mind there was no connection between scientific knowledge and the transformation of the external environment, or, in other words, between science and power.

For Heidegger, technology first became disenchanting with the arrival of the modern age, when it exploited nature for the first time. And what was disenchanting was not so much the machine itself – the steam engine, for example – but a machine way of thinking that allowed nature itself to be approached as something that existed largely to be mechanized. Take the windmill that was introduced into early medieval Europe. As Heidegger argued, the windmill is a device to 'harness' the wind, but, in truth, the contrast between the old and new way of thinking is profound. In the pre-modern world:

> its sails do indeed turn in the wind: they are left entirely to the wind's blowing. But the windmill does not unlock energy from the air currents in order to store it.[42]

Industrialization changed all that, as did modern agriculture. In the pre-modern world fields were planted and crops harvested according to the cycle of the seasons. In the modern era agriculture became a 'mechanized food industry'. This 'enframing' of the world as a 'standing reserve' for regulation and order was what Heidegger found so disenchanting about modern life.

For the salient theme of modern technology was violence. This was a world in which nature is disposed of by man, in which it is a 'resource' to be used. But Heidegger also realized that man himself was now exploited as a resource, or, in his terms, a 'standing reserve'. Man too had become the object, not subject, of technology. For Heidegger this meant more than just reducing people to assembly-line work, to cogs in a wheel. Man could no longer use technology to reveal his own humanity, or as he put it in *The Question of Technology*, 'to experience the call of a more primal truth'.[43] Men had become commodities to be transformed or stored in a way that obliterated human agency. This was the cruel reality of the technological civilization Europe had become and to which Marx gave potent expression in the *Grundrisse*. The machine was no longer a tool manipulated by man; man was manipulated by his tools: 'The workers themselves are cast merely as [the machine's] conscious linkages.'[44]

Disenchantment and Technological Civilization

> Specialists without spirit, sensualists without heart; this nullity imagines it has attained a level of civilization never before achieved.
>
> Max Weber, *The Protestant Ethic and the Spirit of Capitalism*

> Industrial civilization, like the civilization of war, feeds on carrion. Cannon fodder and machine fodder.
>
> Giovanni Papini, 1913, cited in *Origins of Nazi Violence*

Writing in the early 1920s, the American historian Charles Beard became one of the first to speak of a 'technological civilization'. For the West was the first society in history to conceive of the 'civilized' as a way of life defined exclusively in terms of technology. It was the first to associate civilization with the steam engine, and later the internal combustion engine and the aeroplane. It was the first to define civilization as a world which was steel girded, telephonically connected and lit by electric light: in every respect, a machine-created world.

The influence of technology in life grew exponentially. It permeated every aspect of life, both the public and private spheres. It conditioned humanity to think in certain ways, and predisposed it to act in others. It seeped into modern consciousness with results that many people found deeply disenchanting. It even compounded a lack of meaning in people's lives because its language was non-teleological. Nothing can explain why a machine is necessary other than that it promotes a technological 'advance'. The language of the technological age could not refer beyond utility because it could provide no encompassing structure of meaning.

If Beard was the first writer to talk of a 'technological civilization', Agnes Heller defined it in terms I shall use in this chapter. For her, life in the late modern era was dominated by three factors: mediation, efficiency and rationality, all of which disenchanted life in their own way.[45]

Mediation

Mediated by technology, the intercourse between people became more apersonal and functional. Workers were no longer active agents in their own fate but passive actors, bystanders of the future. For what was required of them above all was an ability to deal with stress.

An obscure writer at the time, Hermann Bahr, even divided the whole of human history into three ages: the classical, the romantic and the modern. Each had its own psychological temper: the classical revered reason; the romantic, passion; and the modern? 'When the modern say "Man" it means the nerves.' For Bahr, nervousness constituted 'the release of the modern', precisely because modernity was more stressful than any other age. He even predicted the arrival of a 'new human being'; not Nietzsche's Superman, or the Soviet New Man, but one fully in control of his nerves.[46]

This was especially demanded of the modern soldier. For the accelerating industrialization of war raised stress levels to new heights. What made it so stressful was the length of combat. The average Napoleonic battle lasted one day. But the length of combat increased exponentially in World War I and in some cases was even longer in World War II. In the Pacific campaign the battle for Saipan lasted 28 days and that for Iwo Jima 26.

Such lengthy campaigns produced record cases of post-traumatic stress: 200,000 soldiers in the three main armies on the Western Front, the victims of the 'mind wounds' first diagnosed in the American Civil War. In the Normandy campaign after 60 days of fighting, a US Army study at the time found that all but 2 per cent of surviving soldiers were psychiatric casualties. As a result, fear was not only explicitly acknowledged for the first time, but also taken as a given. Terror openly confessed implied no moral disgrace. The advice was to be found in the US National Research Council's *Psychology of the Fighting Man* (1943). In the same year C. Day Lewis wrote 'An Ode to Fear' which was not literary but literal in defining the symptoms.

> The bones, the stalwart spine
> The legs like bastions
> The *nerves*, the heart's natural combustion
> The head that hives our active thoughts – all pine
> Are quenched, or paralysed
> When fear puts unexpected question
> And makes the heroic body freeze like a beast surprised.[47]

In such a world even courage – the most noble attribute of the soldier – became disenchanted. Before the twentieth century it had meant 'showing nerve'. Now, wrote the French historian Marc Bloch, courage was 'standing under fire, and not trembling'. Wilfred Owen wrote back to his mother from the Front in 1917 that he 'had conquered his nerves'. 'My nerves are under control', an English officer recorded in his diary the year before.[48] If one's nerves took over, then one would suffer shell shock. If that happened, then, like Owen and Siegfried Sassoon, one might be sent off to a sanatorium to be patched up and sent back to the Front.

Efficiency

Industrial civilization also required coordination, which was spread first by the factory bell and whistle, then by individual watches. In

1863 Marx wrote to Engels: 'The clock is the first automatic machine applied to practical purposes. The whole theory of the production of regular motion was developed through it'.[49] Discipline was the key concept. Industrial workers were expected to work with machine-like precision.

In the first years of the twentieth century Frederick Winslow Taylor developed a form of behavioural engineering that treated the body as a machine. In doing so, Taylor objectified the human subject by regarding it not as a person who speaks to another subjectively but as a concrete and desubjectified manifestation of laws revealed by natural abstraction.[50] Taylor forged his ideas and methods in the factories of the north-east United States in the 1870s. Taylorism was the ideal of efficiency applied to production as a scientific method. Its dream was of workers and machines working in synchronized fashion at maximum speed.

It was often said that in his quest for efficiency Taylor did not distinguish between men and machine but, in fact, he saw the new man in terms of Darwinism, as further evidence of evolution. In his treatise 'Shop Management' (1903) he talked of the 'New Man' in almost evolutionary terms. Men would have to adjust to higher rates of speed or perish. The fastest would be the fittest.

> In reaching the final high rate of speed which shall be steadily maintained the broad facts should be realized that men must pass through several distinct phases, rising from one pace of efficiency to another.[51]

In the next stage of its evolution humanity would only survive by being efficient.

As a contemporary writer noted, the separation of the worker from the means of production was one of the historical conditions of modern capitalism, but Taylorism did more: it dissociated the worker from the control of the work process. Taylor's ideal worker was an unthinking one with no intellectual autonomy, capable only of mechanically accomplishing standardized operations – in

Taylor's own words, an 'ox' or 'an intelligent gorilla' (a 'chimpanzee' as Celine put it in *Voyage au bout de la nuit*).[52]

Taylor, in other words, helped instil American capitalism with a fierce obsession with time, order, productivity and efficiency. In his main book *The Principles of Scientific Management*, he famously proclaimed that 'in the past man was first; in the future the system must be first', and it was as a system that modern war too must be seen. Armies were now an industrial workforce governed by the laws of scientific management. The more rational use of 'human *matériel*' was the most effective way to control the chaos of individual wills. According to one World War I German soldier, the most terrible aspect of war was that everything had become mechanical. 'You could almost describe it as an industry specializing in human butchery.'[53]

Rationality

Thirdly, technological civilization was intensely rational. In using that word, of course, we must, like Weber, be careful to distinguish reason from rationality for they are not the same. Reason had always been valued as a mark of civilization, celebrated as the triumph over the irrational or the instinctive. This was the way in which the Europeans explained, to their own satisfaction at least, their superiority over 'savages' who were, by definition, 'unreasonable' or 'irrational'. It was also a quintessential western idea that a society that was technologically advanced was the most civilized. Even in harnessing technology to war the Europeans believed they were being civilized. After all, the more rational a society was, the more reasonable it believed itself to be. And its main object in going to war, at least against savages, was to get them to 'see' reason, through the machine gun.

But in a technological civilization reason and rationality were different. Rationality was a way of looking at the world in which the meaning of an act derived entirely from its utility. Within the framework of practical rationality all means of procuring desired ends are viewed as 'techniques' or 'strategies' rather than as systems of values adhered to on the basis of ethical standards.

The rationalization of modern life included the widespread use of calculation as a strategy of social action; the freeing of social action from all magical thought; the widespread use of technical and procedural reasoning as a way of determining practical outcomes and mastering everyday life. Weber believed that social processes in the West were becoming more and more reliant on technical knowledge and 'calculative reasoning'. And this led to greater utilitarianism in life, to 'the absence of all metaphysics and almost all residues of religious anchorage in the sense of the absence and rejection of all non-utilitarian yardsticks'.[54]

This soon presented the West with a painful dilemma. Weber took his phrase 'the disenchantment of life' from the poet and playwright Schiller. Yet Schiller had also warned that: 'A man can be at odds with himself (and his humanity) in two ways: either as a savage when feeling predominates over principle; or as a barbarian when principle destroys feeling.' As modern men, the Europeans thought they had escaped barbarism through the use of reason. But if they thought that savages were uncivilized for putting feeling first, they soon came to see themselves as savages for sacrificing feeling.

This was the ultimate paradox of an industrial civilization. As Heller tells us, the latter is both liberating and alienating at the same time. It eases life physically, but it alienates spiritually. And nothing was more alienating than serial slaughter. In the case of war it made a nonsense of chivalry, virtue and even courage (courage freely chosen). In Robert Musil's novel *The Man Without Qualities* (his critical inquisition into the violent end of the pre-1914 world), the hero Ulrich abandons his cavalry regiment for a more modern profession: he enrols in a civil engineering course. Musil himself had been a soldier, engineer and mathematician before becoming a novelist, and perhaps Ulrich expresses elements of himself. Interestingly, the hero is alienated from civil engineering too. The world, he complains, 'is simply ridiculous if you look at it from the technical point of view'.[55] He might also have been talking about modern war – that is what made it so disenchanting.

Post-industrial Paradigms

To grasp what had happened we need to look at the paradigm that governed military technology and military thinking in the industrial era. It was one that revolved around the central technology of the industrial revolution: the steam engine. Even earlier war had been dominated by a paradigm centred on another technology: the clock. Both paradigms inspired metaphorical frameworks for understanding science, and through science, the practice of war.

War as force

Although Cicero once compared the regular heavenly motion to the regularities of time pieces, the specific use of the clockwork metaphor for the universe did not really take off until the fourteenth century when the counterpoise clock was invented in Germany and public time pieces began to spread throughout Europe. Subjective time was replaced with the objectification of time: public time. Time was now measured. Time keeping passed into time serving, time accounting and time rationing. Even the Christian Church (in the West, but not significantly in the East), which measured human action by the template of eternity, allowed human life to be contaminated by clockwork time when it installed clocks on church steeples in the major towns. In so doing it subordinated its own liturgical principles to the increasingly materialistic outlook of a society for which time was money.

By the seventeenth century the clock metaphor began to be applied to the human anatomy, as well as the macrocosm of the universe. Descartes, Huygens and Leibnitz all explained nature (the human mind excepted) in terms of the interaction of parts (in this case, matter). Huygens even revolutionized clockwork design by inventing the pendulum and the adjusting spiral. Newton's paradigm of the universe was a clockwork mechanism described in terms of motion, generated by force: the way moving parts of a clockwork mechanism were driven by the pull of weights attached to a rope. Newton transformed the metaphor of clockwork known to his predecessors into something scientifically more tangible: *force.*

The precise measurement of the passing of time was a prerequisite of modern science. The introduction of the timetable – the tabulation of a sequence of events taking place at preordained intervals defined with meticulous accuracy – was the least dramatic but most powerful instrument of change in the whole process of modernization, and that includes the modernization of war. As Manuel de Landa contends in *War in the Age of the Intelligent Machine* (1991), the 'clockwork armies' of the eighteenth century were clockwork mechanisms: the rigid formation of men and weapons was incapable of exerting any individual initiative on the battlefield. A clockwork, as opposed to a motor, only transmits motion from an external source – it cannot produce any motion on its own. In the case of armies, writes de Landa, it was not so much their inability to produce motion that characterizes them as clockwork armies (although they were indeed slow and clumsy to manoeuvre in the field) so much as their inability to process new information (to use data for the ongoing battle and to seize advantage of fleeting tactical opportunities). Individual initiative was limited; the military mechanism, like a clock, was orderly, calibrated and precise.

> In an era when rumour was the fastest method of communication, 250 miles a day compared to 150 miles per day taken by the courier relay systems, the tactical body favoured was the one with the least local initiative, that is, the one that demands a minimum of internal information processing.[56]

But if the clockwork paradigm did not revolutionize the European battlefield, it did give the western Europeans their technological supremacy over the Turks in the intricate relationship between the clockwork universe and drill. Clockwork precision was essential to drill, which was introduced into Europe by Maurice of Nassau in the 1590s. And without drill an army could not march in time.

No such thinking could be found in the Ottoman Empire, but then many European travellers in the seventeenth century were struck by the Turkish aversion to clocks which reflected very different ideas about the measurement of time. The clock encouraged Europeans to abstract and quantify their experience of time, a process of abstraction which led to modern physics. The Turks, by contrast, stuck to the old pre-modern ways of thinking about it. Most of their clocks were imported from Europe and the practice soon arose of sending craftsmen with the gift of clocks to demonstrate their use and repair them when necessary.[57] Inevitably, these attitudes were translated into the military sphere as well. Ottoman soldiers did not march in step but ambled independently. The difference reflected a wider aversion to the close order discipline – or timing – also introduced by Maurice such as the technique of firing in salvo, which involved each rank firing simultaneously at the enemy and then retiring to reload while the other ranks followed suit and created a continuous hail of fire. To perform this manoeuvre required not only fortitude, but perfect coordination. Precision and harmony mirrored the preoccupation of the age with time.

War as energy

If seventeenth-century philosophers saw the universe as a gigantic piece of clockwork, to their nineteenth-century successors it appeared to have many of the attributes of a heat engine. Even the living organism (man) was conceived of and studied as a heat engine burning glucose or starch, fats and proteins into carbon dioxide, water and urea.

The most important of these heat engines, however, was not man but the steam engine, and contemporaries knew it. Steam engines in the early 1800s often had decorative flourishes derived from ancient classical architecture. Even the engine houses and pumping stations sometimes looked like churches. The steam engine was the symbol as well as substance of the age.

In the same way as with the mechanical clock, the steam engine led to another revolution in science. The mathematician Carnot's *Reflections on the Motive Power of Fire* (1824) spelled out the physical principles that constrained the operation of any heat engine. Using Carnot's principles, it was possible to calculate the maximum amount of work that any heat engine could produce. The ultimate limitations were not in the design of any device but inherent in the way that nature worked. Steam engines became society's prime mover and inspired the new science of thermodynamics. Thermodynamic descriptions supplanted Newton's central concept of force with a new feature – *energy*.

Energy, declared Werner Heisenberg in his lectures on physics and philosophy delivered late in life, at the University of St Andrews in the 1950s,

> is indeed the material of which all the elementary particles, all atoms and, therefore, all things in general, are made and at the same time energy is also that which is moved. . . . Energy can be transformed into movement, heat, light. . . . Energy can be regarded as the cause of all changes in the world.[58]

The modern world, wrote Ernst Junger, required 'the growing transformation of life into energy'.[59] War, like the working of the physical world, came to be viewed by the late nineteenth century in terms of energy. Paul Virilio writes of the French revolutionary armies as the first 'motorized' armies in history influenced as they were by the 'energy' of their own revolutionary ideas. The revolutionary song (the *Marseillaise*) was a form of kinetic energy that pushed the masses towards the battlefield and released, in Clausewitz's words, an energy 'that could no longer be checked'.[60] Cannon was not only a new weapon in Napoleon's hands; it embodied the power principle at the heart of his thinking: crush the opposition and inspire fear. For that reason George Meredith called him 'the hugest of engines, (albeit) a much limited man'.[61]

Now it is important to recognize that both changes in cosmology involved concepts of time: they encouraged generals to think of the timing of an attack, the speed of manoeuvre, the tempo of a campaign. Both heightened human interactions in terms of immediacy. The Blitzkrieg – perhaps the most elegant and initially devastating idea of war as pure energy – is a case in point, although it succumbed as all energy does to the thermodynamic principle of entropy. For in the case of the Russian campaign the German army's very success led to what strategists call the 'culminating point' of operations, when forces move too far or too fast to be supported logistically – the point at which tactical success often ends in strategic ruin. In the case of Leningrad in 1941, the tanks arrived at the gates of the city far in advance of the infantry and thus missed the only opportunity to seize it. Instead they were forced to settle down to a 900-day siege which was eventually broken.

Entropy is not a bad way of looking at how industrialized warfare exhausted its own possibilities, especially with the greatest release of energy of all: the atomic bomb. With the atom, as Bernard Brodie remarked early on, the purpose of war became its own negation.

War as information

Now, like the clock and the steam engine before it, the computer has given science a powerful metaphor for ordering the world. Energy as a concept, of course, has not disappeared any more than Newtonian science was 'disproved' in the nineteenth century. What has happened, instead, is that we have a new prism through which to see the world. For the information-processing viewpoint inspired by the computer provides us with a different understanding of life as *information*.

One way in which we can see that the information age has transformed our thinking is the way we now tend to look at Darwinian evolution. In challenging our old industrial, Darwinian styles of evolutionary biology, it has changed perceptions of ourselves, both individually and socially. For as a citizen of the industrial age Darwin came to view living things not as a 'whole' but as the sum total of parts 'assembled' together into a complex machine.

> Almost every part of every organic being is so beautifully related to its complex conditions of life that it seems as improbable that any part should have been suddenly produced perfect as that a complex machine should have been invented by man in a perfect state.[62]

Such remarks only served to confirm Spengler in his observation that Darwin's entire thesis amounted to 'the application of economics to biology'.[63]

Today, we now believe that all organizations are interacting bundles of relationships and all interact with their environment: we call this 'evolution by positive feedback'. Indeed, scientists are beginning to view evolution as the steady advance towards the 'increased complexity of an organization'. Organizational complexity, in turn, is equivalent to the accumulation of information. In other words, writes Jeremy Rifkin, evolution is now seen as an improvement in information processing.[64]

The more successful a species is in processing more complex kinds of information, the better it is able to adjust to a greater array of environmental change. What drives this evolution is increased computational ability. In this 'evolutionary picture', or cybernetic world, we find survival *is* information processing. Our former belief in the survival of the fittest has been replaced by a belief in the survival of the best informed. Life itself goes beyond crude strength to knowledge. The human being is seen as an information processor bound in a complex network with other human beings. Even DNA has an information storage function. Molecules within the cells of living humans contain useful information about the history of speech

(DNA's information storage function alone is reason enough to regard life as in essence an information-processing process).[65]

All models, writes Stephen Toulmin, 'begin as analogy but later become more realistic and closer to strictly the literal "mode of speech"'.[66] Our view of war as information processing (information can alter judgement, morale, it can empower and disempower at the same time) encourages us to think about it in new ways. For even war is being brought under the control of cybernetic principles. Armies are increasingly being thought of as information systems embedded in networks of relationships. The industrial era thought of the whole as an aggregate of assembled parts that made it up. Cybernetics, by contrast, views the whole as an integrated system.

The rise in the speed of information-processing capabilities constitutes a radical paradigm – the rise of cybernetic theory. Norbert Wiener, the man who first popularized it, defined information as the

> name for the content for what is exchanged with the outer world as we adjust to it and make our adjustments felt upon it. The process of receiving and of using information is the process of our adjusting to the contingencies of the outer environment and of our living effectively within that environment.[67]

Information, then, consists of countless messages and instructions that go back and forth between things and their environment.

Cybernetics, in turn, is the theory of the way those messages or pieces of information interact with one another to produce predictable forms of action. The 'steering' mechanism is negative feedback, which provides information to the machine on its performance so that it can attempt to close the gap between what it does and what is expected of it. The result is information processing. With the aid of the computer, cybernetics has become the foremost organizing principle of our world. In turn, the computer has ensured the institutionalization of cybernetic principles as the central organizing mode of the future.[68]

We now inhabit a network world which we share with computers. Our humanity has been redefined for us. To be a genius is to have greater situational awareness. To be knowledgeable is to be better informed. To be better informed is to know of every change in circumstances in the external environment. Wisdom has largely become a matter of information processing. And knowledge is no longer the discovery of facts as in the past but an ongoing process of data evaluation that, as subjective beings, we evaluate with machines.

To quote one of the most important US military publications, the *Army After Next Annual Report*:

> Knowledge is paramount . . . the unprecedented level of battle space awareness that is expected to be available will significantly reduce both fog and friction. Knowledge will shape the battle space and create conditions for success. It will permit . . . distributed, decentralized, noncontiguous operations. . . . It will provide security and reduce risk.[69]

The reduction of risk is one of the key objectives of the new technologies, especially risk to our own soldiers.

Now it is this change which constitutes the re-enchantment of war, and not only because it is more humane. If the network has intrinsic value, so too do the soldiers who are part of it. Similarly, casualties can be kept to a minimum; cities don't need to be dismantled, only shut down; armies don't need to be taken out, merely neutralized – in each case not for very long.

It is because everything is interrelated that soldiers are increasingly regarded as information gatherers, linked into a larger network which processes the information they both provide and are supplied with. Being able to respond quickly to a fast-changing environment is now the key to success. And success is no longer measured according to the ability to kill in large numbers, or to level cities, or immobilize armies in the field – a kind of crude social Darwinist struggle of the fittest which reached its nadir in battles such as

Verdun. Instead, success is measured by the ability to absorb an increasing amount of information, and to direct firepower at certain nodal points. The Americans are aiming to produce a single 'system of systems' that will connect space-based, ground-based and air-based sensors and decision-assistance technology, and thus provide the 'information superiority' that will allow a commander to prevail with the minimum of material harm to himself or others.

Once again the new cosmological understanding of war is about time: immediacy. But with one essential difference: time and speed have become conflated. For targeting is now measured by the nano-second. An error of a billionth of a second means an error of about a foot, the distance light travels in that time. One nanosecond equals one foot, an equation that can make all the difference between taking out a target with or without collateral damage. Knowing the exact time is an essential feature of delivering airborne explosives to specific buildings, or even rooms in a building, and thus minimizing loss of life for belligerents and non-belligerents alike.

Metaphors of war

Clock	*Steam engine*	*Computer*
• Dominant tool in society. • Tool as metaphor for science leading to the new science of Newtonian mechanism. • Metaphor for scientific worldview based on *force*. • War as directed mechanized force.	• Dominant tool in society. • The object of scientific study leading to the new study of thermodynamics. • Metaphor for scientific worldview based on *energy*. • War as directed energy.	• Dominant tool in society. • Tool for science and object of scientific study leading to new science of physics of computation. • Metaphor for scientific worldview based on *information*. • War as negative information feedback.

Post-industrial Civilization

The industrial basis of victory was clear in World War I. It was even clearer in World War II. The poet Archibald Macleish was more precise than perhaps he intended when, in an article in the *Atlantic Monthly* in 1949, he claimed that the United States had 'engineered' a brilliant victory.[70] But it was, of course, a victory at high cost: the carpet bombing of German and Japanese cities, a process that left both societies in ruin. So it is not altogether surprising that it was in the course of World War II that the first use of computers emerged as well as the development of cybernetic theory, both of which have had a dramatic impact on our thinking about war. Subsequently, they have allowed the United States to abandon the old operational concepts such as massed force and sequential operations in favour of massed effects and simultaneous operations.

And this, in turn, owes much to two great developments that in the course of the twentieth century changed the face of war: (1) the speed of communications and (2) the speed of computerization, both of which have ushered in the era of post-industrial warfare. Just as the clock emphasized the importance of timetables and co-ordinated drills, and the steam engine the importance of thermodynamic principles on the battlefield, so the computer has reinforced the critical importance of speed in information processing. For the information age has truly transformed war with respect to the speed of communications. In World War I the telegraph enabled commanders to receive instantaneously from, and give orders instantaneously to, battalions in the field. The fastest transmitter of data in its day, it was capable of sending 30 words a minute. Teletype increased this rate to 66 words a minute in the closing years of the Vietnam War. The computers used in Operation Desert Storm (1991), by comparison, were able to process 192,000 bits of information per minute. In the near future we can look forward to the processing of millions – perhaps even a trillion – bits of information per minute as computer chips become ever more sophisticated.[71]

Even when our current silicon systems reach their physical limitations, which may be sooner than imagined (perhaps within 20 years), new technologies based on molecular and physical sciences, and above all quantum physics, may take us into another age. Nanotechnology has the potential to provide much greater processing capacity by engineering more into the same space, and biological processing systems may increase processing speed. The greatest leap of all may come from quantum technology, with new algorithms based on quantum principles – although at the moment it is far too early to make any firm predictions.

The information age has been vital in another respect: connectivity. Connectivity between computer systems has increased exponentially through digitalizing the separate systems that preceded it: the telephone, radio, television, satellites and, most recently, optical fibres. All of these technologies have been brought together in a network. Integrated digitally, a uniform digital bit stream has replaced the previously separate technology of telephony, broadcasting and data communications. Within 20 years all these systems may be 'on line' all the time, offering military possibilities not yet dreamed of which go far beyond the 'full spectrum dominance' that the US armed forces are aiming to achieve by 2010.

Secondly, computerization has helped produce a network which enables machines to process information digitally. The first-generation model marketed after 1951 had been vacuum-tube based. Beginning in 1960, IBM produced large transistorized computers, second-generation machines, which launched the computer age in its present form. The next stage was the present-day microprocessor chip, which was able to perform millions of logical operations per second. In 1972 Intel produced the 8,080 chip, a single piece of silicon containing 4,800 transistors. It was the first true microprocessor. Since then chips have been increasing both in speed and capacity exponentially from 1K to 256M-bit chip. What is the limit – 100 million transistors in a single microprocessor? Physicists expect to go much further than that.[72]

An information society is characterized by the integration of information technology as the key factor in all kinds of production. A growing part of the US military spends its time processing information. Its armies are now totally dependent on computers and Global Positioning Systems (GPS). If they fail, or if the computer crashes, the plans crash with them as well. More and more military activities consist of processing various forms of information. Information is now the driving force of war.

What is truly revolutionary, however, is the manner in which the new technologies have fundamentally transformed the way we think about war, in particular the targeting of *enemies*.

Optimization

The industrial-based technology of World War II resulted in the maximization of casualties largely because of the inaccuracy of area-killing weapons. One example was the 1,000 bomber raids that were unleashed against German cities after 1942, and which Vonnegut experienced at first hand a few years later. The weapon of ultimate victory was the B-17 Flying Fortress, which according to Allied propaganda could hit a 25-foot circle from 20,000 feet. It couldn't, which is why massed attacks resulted in massive loss of life. By the end of the conflict 22,000 of these bombers had been destroyed, with the loss of 110,000 airmen. The losses were attributable, ironically, to the plane's inaccuracy. 'Precision bombing' became a sardonic oxymoron relished with a sense of black humour by bomber crews who knew they stood a good chance of dying because of the imprecision of their machines. As Randall Jarrell put it in his poem 'Losses', the bodies of the aircrew lay among the people they had killed but never seen.[73]

As for civilian losses, they were equally horrendous: 131 German cities and towns were targeted by Allied bombs, and a good number almost entirely flattened. Stig Dagerman, a Swedish journalist who reported from Germany after the war, told his readers that the masses of white faces among the survivors of the bombing of Hamburg reminded him of the faces of fish coming up to the surface to snatch a breath of air.[74]

Today, technology aims at optimization rather than maximization.[75] Bigger is not always better. Optimum not maximum firepower is now the rule. Information equates with safety. It also equates with lethality, and lethality with increasing precision. The number of aircraft now needed to destroy targets has been reduced significantly. In the Gulf War of 1991, one F-117 Stealth Bomber with laser-guided bombs destroyed the same targets that would have required 1,500 B-17 missions in World War II. Even so, the US Air Force (USAF) had to deploy 10 aircraft to be sure of taking out a single target. In Afghanistan 10 years later it was able to budget two targets per aircraft.

Also in the first Gulf War the F-16 and F/A-18 fighters were able to place 50 per cent of their bombs within 30 feet of their aim points, although the smart bombs routinely hit within 3 feet of their targets. Desert Storm was a vindication of the old concept of precision bombing: the central difference was that technology finally caught up with the doctrine.

Cultivation

War has seen another change in thinking which parallels the post-industrial attitude to technology in general. We are much more interested in technology as cultivation rather than manipulation.

In the past societies exploited or manipulated their external environment without thought to the consequences to the environment or themselves. The industrial revolution permitted the ruthless exploitation of nature. In the 1920s an Italian writer, Romano Guardino, deeply distressed at the environmental degradation he had witnessed, published his reflections in *Letters from Lake Como* (1926). Technology, he complained, had moved from understanding nature to exploiting it:

> Materials and forces are harnessed, unleashed, burst open, altered and directed at will. There is no feeling for what is organically possible or tolerable in any living sense. No sense of natural proportion determines the approach.[76]

What Guardino recognized was that it is impossible to make one's self the 'master of nature' if one is part of it, and yet dependent on it at the same time.

Instead of exploiting nature we are now trying to cultivate it. As Lewis Mumford wrote many years ago, the future of technology would lie not in mass exploitation but in the use of machines on a human scale, 'to fulfil with more exquisite adaptation, with a higher refinement of skill, the human needs that are to be served'.[77] And it is increasingly the case in the western world that companies are paying more attention than ever to internal biological rhythms and ecosystem restraints. Regenerative farming with full attention to the needs of the land is the natural alternative to the energy-intensive, resource-depleting, variety-threatening, pollution-producing agribusiness of the modern era which philosophers like Heidegger so disparaged.[78]

The same is true of war. Indeed, 'The Marine Corps After Next' (MCAN) Branch of the Marine Corps Warfighting Laboratory is exploring what it calls a 'biological systems inspiration' for future warfighting. According to its website:

> For the last three centuries we have approached war as a Newtonian system. That is, mechanical and ordered. In fact, it is probably not. The more likely model is a complex system that is open ended, parallel and very sensitive to initial conditions and continued 'inputs'. Those inputs are the 'fortunes of war'. If we assume that war remain a complex and minimally predictable event, the structures and tactics we employ will enjoy success if they have the following operational characteristics:
>
> Dispersed
> Autonomous
> Adaptable
> Small.[79]

The characteristics of an adaptable, complex system closely parallel biology. To deal with the biological is to do least damage to the environment, understood as the social, political as well as ecological context within which war is fought. When the term 'ecology' was first coined in the 1860s it described the holistic study of living systems interacting with their environment. Ecologists look at communities of organisms, patterns of life, natural cycles and demographic changes. And this is precisely what a new generation of the military is doing as well.

Let us go back to Guardino's complaint that in exploiting nature 'no sense of natural proportion determines the approach'. Today this is no longer the case. What is important for victory is what we want from it: stability, durability and the sustainability of a society in peace. And this is all the more important when one country has the means, as showed in the second Gulf War of 2003, to effectively eliminate another's entire leadership, public administration and justice system in a matter of weeks. The aim of war is increasingly designed to preserve as much of a society as possible as well as to preserve the human habitat that enhances the quality of life and thus makes life worth living.

In the run-up to the war the US Army War College was asked to review possible models for its prosecution. Traditionally, warfare unfolds through four stages: 'deterrence and engagement'; 'seize the initiative'; 'decisive operations'; and 'post-conflict'. Reality is never quite that neatly divided, but the College report stressed that phase 4, 'post-conflict', had to start before phase 3, 'decisive operations', or before the war itself.

Indeed, it is useful to remind ourselves that at the end of the war the campaign had achieved most of its objectives: the infrastructure had been preserved, a humanitarian disaster had failed to materialize, society remained largely intact. Unfortunately, the massive looting that followed the end of the major warfighting caused more civil and economic damage than the war itself. The US Army also allowed the situation to get out of control from the start. The fact that the third phase went very well and the fourth very badly only highlighted the length of time it takes for a new paradigm to establish itself in the military mind.

Differentiation

Post-industrial thinking of the sort we are considering has also reduced enemy casualties by focusing on differentiation rather than centralization. The network has replaced the centralized mass system of two world wars. As the US military evolves, it is gradually abandoning its old operational concepts such as massed force in favour of massed effects. Instead of highly centralized attritional warfare it now favours network-centric warfare. According to *Joint Vision 2010* (the most important expression of the official American vision of future war – what might be called its 'conceptual template' for the future), this is made entirely possible by technology.

> Increased dispersion and mobility are possible offensively because each platform or industrial war fighter carries higher lethality and has greater reach. Defensively, dispersion and higher tempo complicate enemy targeting and reduce the effectiveness of area attack.[80]

The Pentagon's aim is to mould weapons systems and peoples into a network-centric style of warfare greater than the sum of its parts. Among other things the system should make it easy to track and attack military targets and provide a command structure that is more resilient and damage proof. The military's resounding success in Afghanistan where units from different branches of the service worked in unprecedented unison has led to a consensus that this is the way of the future. It embodies an impressive change in institutional culture, though one that still has a long way to go. In the Afghanistan War of 2002, stories of Special Forces calling in air strikes with laser pointers may have made the media headlines, but behind the scenes commanders had to queue up for satellite uplinks and bickering broke out over who would get access to Unmanned Aerial Vehicles (UAVs) such as Global Hawks and Predators.

Still, the future is clear. In the first Gulf War commanders took reports by radio and scribbled down troop positions with grease pencils on a map. Now troop deployments are displayed on digital screens. In the most advanced US divisions this wireless Internet system is installed on nearly every vehicle. The aim is to create a dense mesh that can encompass a battle zone and provide troops with far more reliable connectivity than anything that is available today so that they may engage in far more precise targeting and less damage to the society that is being targeted.

Conclusion

> *Premises of the machine age*: – the press, the machine, the railway, the telegraph are premises whose thousand-year conclusion no one has yet dared to draw.
>
> Friedrich Nietzsche, *Human, All Too Human*, p. 278

In William Gibson's novels one of his characters has a remarkable gift. Jacked into his computer he can navigate vast seas of digitalized information and discover the cyberworld's 'nodal points'. He can get an interpretative fix on where a person is spatially and spiritually, and where he will go next. In history, alas, there are no nodal points. All predictions are merely extrapolations of present trends and not all trends lead to the future, or follow their own logic to a logical (or illogical) conclusion. In this book all that I have tried to do is to follow through on the leads which contemporary developments offer us, as does contemporary technological research. If disenchantment arose from the extent to which modernity, in relieving the imposition on our bodies, did so at the expense of our souls, re-enchantment lies in putting us back in touch with our humanity.

The essence of technology, as Heidegger reminds us, is not technological; technology is not independent of history, it is explained by history and, in turn, explains it. The technologies we are now developing have been invented to meet a need, the need to re-enchant war by humanizing it. Whether we can – or should – do so will be the ethical question of the future. It is addressed to the body in pain, to the *biological* principle that will animate the face of battle in the twenty-first century rather than the industrial demands for mass destruction, with their mechanistic burdens on the soldier, which dominated military thinking in the late modern era.

These trends, it is true, were anticipated long before the information age. But writers in the 1920s such as H. G. Wells believed that the very attempt to humanize war would lead to its own negation. In an essay entitled 'Changes in the Arts of War', he accused the western powers of wanting 'nice wars', wars that were honourable, comfortable and dignified. He claimed that they wanted to ban weapons systems they considered 'unsportsmanlike' such as aeroplanes and submarines, and to keep war's horrors within the bounds of human endurance. While finding the attempt laughable, he had one hope: that 'the improvement of war may be synonymous with the ending of war'.[81]

But Wells and his contemporaries missed the point, not surprisingly given the limited technological options of the day. War has not been renounced, it has been 're-enchanted', and the reason is pretty clear. Technology has allowed us to re-engage with our humanity in three respects, one of which has been the theme of this chapter. *Instrumentally*, war is now fought on a human scale with a deliberate attempt to limit destructiveness to society, to the state, to the environment, to the human habitat: to everything, in fact, that makes life worth living. It remains to ask whether the same is true of the remaining dimensions of war, the existential and metaphysical, which I shall discuss in the chapters that follow.

2

The Warrior of the Future: Memes or Genes?

Martin Heidegger tells us that things reveal themselves to consciousness only through the frustration they cause: breaking, or disappearing, or behaving out of character, or otherwise belying their true nature.[1] For the warrior, even more than the conscripted soldier, war had become immensely alienating because its reality was apparently shattered on the Western Front. We are all alienated, of course, to the extent that we are externalized from ourselves by the objectifying process of our activity, not as individuals but as social agents. What was special about the warrior's alienation in the late modern era was the application of technology to war in the form of mustard gas attacks and artillery barrages, and more directly still the extent to which he was subsumed within a technological dynamic that totally instrumentalized the traditional warrior ethos by robbing war of its existential appeal. The West was the first culture that became aware of war not as something innate to man but as something problematic in human existence, something without value outside the narrow framework of utility. In *devaluing* the warrior, it further compounded its disenchantment.

The Disenchanted Warrior

No war was more disenchanting than World War I. It has been said that Wilfred Owen saw the remnants of another age, that of Arthur's knights, in the carnage of the Western Front and heard the music 'in the screaming funnel of a hospital barge':

> And that long lamentation made him wise
> How unto Avalon, in agony
> Kings passed in the dark barge which Merlin dreamed . . .[2]

These lines were influenced by Tennyson's 'The Passing of King Arthur', where we learn of 'an agony of lamentation, like a wind that shrills all night in a waste land'. And that lamentation, in turn, was taken up after the Great War by T. S. Eliot to personify the war itself, a wasteland in which even a warrior could find no existential appeal.[3]

If Weber's concept of detachment helps us understand what made war disenchanting, the concept of alienation helps us understand what made the warrior himself disenchanted with his profession. In *The Phenomenology of Mind*, Hegel uses the term 'estrangement'. By this he means a historical sweep in which man strives to realize himself in history (usually unsuccessfully) by stages which gradually reduce his estrangement. Hegel was the first philosopher to understand that human beings can experience their own activity as something inauthentic, something that is estranged from the 'self'. He described the experience as 'alienation'.[4]

In one sense, of course, alienation is inevitable in that it separates the essence of humanity from the reality of existence. Man, after all, is alienated from his nature by the acceptance of limits placed upon him by nature itself, both in terms of his relationship with it and the limits imposed by nature on his abilities, physical and mental. In that sense, alienation is inescapable: it is a radical dissociation of self into a subject that strives to control its fate and an object that finds itself manipulated by others. For Hegel this was really a natural condition, for man lives in society: he is not only an 'I' but also a 'me' (he is what others think of him). Our very identity is shaped not just by our own will but by the will of others.

In time, however, alienation became associated as a condition of modernity itself, especially in its industrial manifestation, the one to which Ruskin objected most. In 'The American Scholar', Emerson underlined his own anxiety about the advanced division of labour that was fragmenting human integrity. His fear was that man would be subordinated to his own creations. 'The state of society is one in which the members have suffered amputation from the trunk and strut about so many walking monsters – a good finger, a neck, a stomach, an elbow but never a man. Man is thus metamorphosed into a thing, into many things. . . . The priest becomes a form; the attorney a statute-book; the mechanic a machine.'[5]

The warrior too found himself estranged from his craft as an activity. The first war in which this was probably true was the American Civil War (1861–5). The Union army was the first industrialized army in history and the American Civil War offered the first industrial battlefield, one memorialized in Stephen Crane's *The Red Badge of Courage*. In the book battle itself is repeatedly called a 'vast machine' that produces only corpses. The main theme of the novel is the hero's outrage against his own impotence, which provides the main motivation for acting, both when he runs away and when he fights. His final feats of heroism are spurred by his resentment after overhearing a conversation between two officers in which they refer to soldiers such as himself as 'mule-drivers,' not men, but pack horses in an industrial army. Even the industrial battlefield makes a mockery of his warlike dreams, 'the Greek-like Homeric struggles' that had inspired him in the past.[6]

There was nothing Homeric about the industrialized killing ground of either the Somme or Iwo Jima. As a US marine said of the Okinawa campaign (1945), it was common for replacements to be killed before their comrades even knew their names. 'They were forlorn figures coming up to the meat grinder and going right back out of it like homeless waifs, unknown and faceless to us, like unread books on a shelf.'[7] The very scale of death on the battlefields of the two world wars disenchanted not just war, but the warrior ethos, especially as it had evolved in the western world since the Greeks. What it did was to prompt society to ask a question as old as war itself: are warriors born or made?

The Warrior Meme

The biotechnological age we have entered induces us to ask a similar question. Does a person's genetic makeup predispose him or her to a particular type of human behaviour? Is his or her chosen trade a preference or a biological disposition? Does it demand innate abilities or acquired skills? Is it a product of genes or memes?

Richard Dawkins has popularized the idea that human development is mimetic as well as genetic. The word 'meme' was first coined in 1976 to describe the practices, conventions and taboos that codify genetic choices. Unlike a gene, which is copied to the offspring from its parents, a meme by analogy is anything that replicates itself or is copied via memory. Genes are instructions for making proteins, stored in cells of the body and passed on in reproduction. Their competition drives the evolution of the biological world. Memes, by contrast, are instructions for carrying out behaviour, stored in the memory and passed on by imitation. Their competition drives the evolution of consciousness and thus culture.[8]

A meme can be anything from a good idea to a poem, anything, in other words, that spreads by imitation. In the case of the warrior ethos, the western meme was the example of the great warriors in the *Iliad*. Memes exist in a meme pool – like genes, they evolve. They are copied, altered and selected at different times. In the case of war, we call this the evolution of the warrior ethos. Every era rewrote the Achillean legend for its own needs.

This was quite literally true in the case of Alexander the Great, who carried the shield of Achilles 11,000 miles to India where it saved his life. Alexander, of course, thought that he was genetically descended from Achilles via his mother's line, but it was mimetic repetition which was of historical importance. At the height of the battle of Granicus, he turned the battlefield into a Homeric scene of single combat when he galloped out in front of his men and struck the Persian commander in the face with a spear. The fact that Homer's heroes were essentially armed with the same weapons as Alexander's hoplites made identifying with them all the more natural, and yet coming to terms with the phalanx and the unheroic infantry tactics of the day all the more difficult.[9]

Now, we must remember that memes, like genes, adhere to the rules of Darwinian selection. They spread because they are good at spreading. The first rule of the replication of memes is that replication is not necessarily for the *good* of anything. In other words, there is no necessary connection between a meme's reproductive power (i.e. its fitness from its own point of view) and its contribution to our fitness. Yet whether we see the warrior as a hero or a parasite, the warrior meme is a powerful one, which is why it has endured for almost three millennia. Indeed, it is expressed in three distinctive forms determined by different historical eras:

- pre-modern 'excellence';
- early modern 'merit';
- late modern 'authenticity'.

Today it can no longer be invested with the same weight and significance. For with the biotechnological revolution the old distinction between nature and nurture is becoming blurred. Genetics is becoming far more important than mimetics. Memes are giving way to genes. To put it crudely, if memes were designed (or programmed) to breed the warrior, genetic manipulation now promises the possibility of breeding out all the imperfections even the most heroic warriors have displayed through history – for the first time it offers the warrior a chance to *breed true*.

Warrior meme 1: Excellence

From the time of Hesiod the mutual aggression governing the animal world had been contrasted with the orderly social world of men. Two centuries later Alcmaeon of Croton argued that what distinguished humans from other animals was the diversity of their *social* skills. The Trojan war heroes show a diversity of human types: the sullen defiance of Ajax, the mercurial heroism of Patroclus, the cleverness of Odysseus. It is this diversity which makes society possible: the cooperation of men with different trades and different talents.

This was still a civilization, of course, that was closer to animals in its folk memories than we are today. Indeed, in Homer's epic the heroes have a marked affinity with animals. The Homeric hero is 'like a lion'. Diomedes's distinctive violence is compared to that of a lion breaking the neck of a bull calf. The mere sight of Hector like 'a fine maned lion' is enough to terrify and scatter the Danaans. Even Patroclus, though destined for death, fights to the end, as a lion would do, showing both the animal's courage and its ferocity in combat, qualities that are innate in all lions but not, of course, in all men.[10]

The Greeks were fascinated by the fact that if all lions were courageous and one hare as cowardly as the next, not all men were lions, or for that matter hares. But the courage of a lion, like the cowardice of the hare, is the presence of a need. Need is an animal's immediate experience. The radical difference between humans and the rest of the animal kingdom is that we are not creatures of our immediate needs. Unlike the lion, the warrior, as often as not, is prepared to risk his life for prestige rather than a material reward. He has to in the *Iliad* if his gift of excellence is to win the admiration of friends and foes alike. What is important is that the warrior is born and reveals his natural abilities in battle. It is impossible to rise above one's nature; *arête* or excellence is realizing what is innate to oneself.

The hero demonstrates *who* he is, not what he is. And the 'who' is measured against others (Achilles against Hector), for it is through competition that excellence is demonstrated. It is not to be found necessarily in victory; it can also be found in defeat. And it is not measured instrumentally by motivation (Achilles's petulance in no way derogates from his greatness, which comes from his innate abilities and his willingness to test them in battle).

The concept of excellence, therefore, is the bringing into the light of that which constitutes its nature. The Greeks believed that we are what nature makes us: brave or cowardly, resolute or irresolute, decisive or indecisive. Thus in the *Nicomachean Ethics* Aristotle begins with the observation that what we find excellent in a flute player, sculptor or any artist resides in the function he performs. The artist is born and his life is one in which his talent is fulfilled – or unfulfilled. And while there is room for perfecting one's talent (even the best flautist must practise), excellence is intrinsic, not acquired.[11]

Although the warrior practises his 'art', although he seeks out opportunities to demonstrate his *arête*, he is also bound by what nature permits. In that sense his genetic programming is not what we understand by genetics today. For it is only by recognizing the limits to our nature, i.e. that our nature has limits, that we can live in harmony with ourselves as well as others. It is the limits that we acknowledge – or should acknowledge – that determines the character of means. For the Greeks illimitability was a synonym of evil. The end must define whatever serves as its means – a too courageous warrior (such as Achilles's close friend Patroclus) will put himself and others at unnecessary risk. In Greek thought the concept of limit readily passed into that of the mean. And the mean best calculated to realize a particular end came to be viewed as in itself the best. The moral excellence of the warrior, therefore, is not only courage but the cultivation of the mean of passion, which in the soldier lies between two extremes: foolhardiness and cowardice. In limiting ambition, the concept of limit makes for moderation.

The pursuit of excellence marks the birth of humanism – an exclusively human realm in which humans compete with one another, not with the gods. But the gods, of course, have one advantage over humans: immortality. Human beings live in the world of history, and they live only once. Everything that owes existence to human action, such as artworks and words, is perishable because of the mortality of the authors. Only in reporting heroic deeds can actions, uprooted from the sphere of the perishable, be made equal to the acts of the gods.[12] In the world of the *Iliad* the poet's function was to sing the deeds of the warriors, a privileged caste with its own status. In a warrior society in which all were ostensibly equal, inequality was based on praise or blame. The world of the heroes was one in which the chief good was to be *spoken of* respectfully and to win that respect through one's deeds. To preserve his reputation was the goal of every hero.

By the fifth century and Sophocles's play *Ajax* something had changed. During the funeral games for Achilles his possessions are distributed among his comrades in arms. Ajax is furious because the armour of Achilles is awarded to Odysseus, not to him. This act is taken to dishonour him in the eyes of the assembly. His status is on the line. Nursing his injured pride, he decides to avenge himself. But the goddess Athena (Odysseus's protector) visits him with madness and he ends up wreaking vengeance upon a flock of sheep, which he mistakes for his fellow Greeks. When he comes to his senses, pride insists that he take his own life.

The key word in Sophocles's play is *atimos* – humiliation. Ajax is humiliated by turning his vengeance on animals, not men. When he recovers from his madness he has no regret for his actions. There is no apology for his deeds, only an overwhelming frustration, a belief that he has been penalized by divine intervention. He is, in that sense, a warrior with no real subjectivity as the Greeks by the fifth century understood the term. Athena says the gods love the wise but hate the base (*kakoi*). It is an extraordinary choice of words to employ, for *kakos* means base or cowardly. It is an astonishing term to apply to a man who is the supreme exemplar of heroic excellence, a man who, we are told by Homer, is after Achilles 'the best of the Achaeans'. It can only suggest that the excellence of Ajax is less than meets the eye because of his lack of *self-knowledge*, the mark of a truly wise man.[13]

What the play reveals is that if the Greeks of Sophocles's day were still preoccupied with personal prestige, the ideas that made heroes truly heroic had passed by the fifth century. Ajax is presented to us as the last of the heroes. His death is, in fact, the death of the old Homeric ethos as the true warrior became a man whose pursuit of excellence must be not only for its own sake but also for that of society. When asked to define courage, the Athenian general Laches cites the example of the man who stands at his post and faces the enemy, an image that calls to mind the hoplite standing in line, not the heroes of the *Iliad* yearning for individual glory. The shield of Achilles was forged for the individual warrior. By the fifth century it was the means by which the phalanx held together. Plutarch put it directly: 'men wear their helmets and breastplates for their own needs but they carry their shields for the men of the entire line'.[14] The Greeks regarded the casting away of shields as a form of cowardice for it opened up others to attack. Hence the well-known Spartan mothers' instruction to their sons to come back either bearing their shields or on them. In the new war ethic there is no place for individual heroism when this means breaking ranks. Excellence now lay less in single combat than in discipline.

It is a mark of this change that the warrior meme is now appropriated by the public. The warrior is no longer above language or discussion. He is subject to criticism, or what Jean-Pierre Vernant calls 'a rendering of accounts'.[15] He is now part of a social dialectic. And it is only through 'cooperative' as opposed to 'competitive' excellence that he can put himself beyond reproach.[16]

There is no better example of this in classical literature than the figure of Aeneas in Virgil's epic poem, a hero who differs from Achilles in two vital respects that illustrate what had happened to excellence in the centuries since Homer. First, he has *humanitas*.

What distinguished the Romans from the barbarians, in their own eyes at least, was that the latter exhibited *feritas* (a warlike nature made up of conflicting emotions, pride and anger). It is this insight which makes the Virgilian hero different from the Homeric. For the hero must struggle not only against the external world but also against the savage elements within himself. To show excellence is to struggle against one's baser nature. In other words, Aeneas has a psychological dimension that the warriors of the Mycenaean age lack. He struggles with himself as much as with the enemy, and in rising above his baser nature he shows the piety expected of a Roman soldier.

This is the second change. '*Vicit iter durum pietas* – piety has conquered this hard road'.[17] Unlike Achilles, Aeneas has a vocation, one which appears in the double character of a duty as well as a desire. And Virgil did justice to both. To follow a vocation does not mean happiness, but once it has been learned there is no happiness for those who do not follow their destiny. Piety is to accept one's destiny – and not go beyond it. As André Malraux famously declared, 'When man faces destiny, destiny ends and man comes into his own'.[18]

But piety is in the end not enough. Despite his many virtues Aeneas has little imagination. His high sense of duty is informed by no real love of goodness for its own sake. His piety is wanting in simple humanity. For love of Rome he will always be prepared to sacrifice his happiness, and often the happiness of others. For he is a symbol of a people who came to conceive of a great idea, *humanitas*, but who were not humane enough to realize that the idea is a mere abstraction unless humanity is real.

Aeneas's piety strikes us as hollow. That hollowness was expressed in the closing centuries of the Roman Empire even by a philosopher-king like Marcus Aurelius, who could hardly have set his sights much lower. What else can we conclude from his observation that the activities of men are 'smoke and nothingness'; or his belief that the clash of two armies is nothing but a clash of 'puppies over a bone'; or his description of his own Samartian triumph (celebrated in 176 AD) as the self-satisfaction of a spider who has caught a fly? This is not merely empty rhetoric. Marcus Aurelius means it. He's in deadly earnest. The significance here is not that man's activities are unimportant but that they're not quite real. And it's the unreality of the world – the real world – that appears in Marcus Aurelius's observations on human life.[19]

There was by this time a sense of another world: that of God, a world that Marcel Gauchet attributes to the internal contradictions of empire. For once a state emerged with a logic of expansion, the concept of the universal 'burst upon the scene of human experience'. Once the state reordered the world (or creation), how could the individual gods who constituted that world (its ethnic deities) survive? Monotheism was built into the system. The state had to legitimize its power. In conquering the world, it had to reorder it. There was a dynamic at work here towards decentring customary life and customary convention and redefining both in terms of a universal ideal, the empire of a single God. Here was a rift between immediate reality and ultimate truth. The triumph of Christianity was foreshadowed in this logic. This, adds Gauchet, does not explain why Roman concerns yielded to Christian convictions, but it does make their convergence less implausible.[20]

It is this change which leads us to the Middle Ages and the further evolution of the warrior meme – 'excellence' as sacralization, proving one's excellence in the eyes of God. Medieval Europe invented the crusade, and in so doing produced a new type of warrior, one defined by the asceticism of his calling. Jean de Bueil, a good witness of his own times, saw war as first of all a school of asceticism. It required effort (pain and travail), it required hardship (fasting and the wearing of armour), all for one end: acquiring 'a perfect glory in this world'.[21] Glory, however, did not come without a cost. For we now know that the crusades were not like nineteenth-century colonialism – a form of outdoor relief for the impoverished younger sons of minor families seeking a reward in this life, not the next. Far from it. They were ruinously expensive and justified only by the belief that a warrior's true reward was the atonement of sin. The financial sacrifice of crusading was an essential element in a culture of piety. By the end of the twelfth century a crusader was someone who had taken the cross to go to Jerusalem, accepted obligations of a specific type of pilgrimage and could expect to be away from home for at least three years. The commitment could claim up to four times his annual income. Knights usually financed their journey by mortgaging their property and often, in effect, their families as well.[22]

And there was also another element of piety here that is worth emphasizing. Jonathan Riley-Smith, who has done more than any other writer to explain the crusades as an act of Christian devotion, reminds us of the depth of the crusading commitment. For there was in the crusading soldier the idea of the defence of a wider Christian community. The supranational nature of the crusades was one of its distinctive features. A crusading army was always treated as representative of all Christendom. Love of Eastern Christians in the territories occupied by Muslims was a chief motivating factor in deciding to join up. It was, of course, a one-dimensional world for there was no echo here of Christ's injunction to love one's enemy, still less of St Augustine's insistence that it was more virtuous to love one's enemies than one's friends. Love of one's neighbours meant love of one's fellow kinsmen. This was a narrow view of humanity, indeed a familial one.[23]

Here, writes Norbert Elias in *The Civilizing Process*, was no longer the direct pleasure of the hunt or the blood lust of battle. Instead, the pleasure lay in the closeness to one's friends and enthusiasm for a just cause.[24] This was the ideal of a Christian fellowship permeated by Christian values. The words most commonly used to refer to fellow Christians were *fratres* (brothers) and *amici* (friends), and this at a time when the word *amici* as often as not meant kinsmen. The use of such terms would have conjured up the idea of a family at risk. In this respect, the crusades can be seen as essentially a blood feud.[25]

Warrior meme 2: Merit

Unamuno, the great Cervantes scholar, chose as the most beautiful passage in *Don Quixote* a moment in the second volume when the knight and Sancho Panza come upon some peasants carrying bas-relief carvings from an altar decoration. Concentrating on the images of St George, St Martin and St Paul, the Don is moved to state the difference between the saints and himself: 'they . . . fought in God's wars while I am a sinner and fight in humanity's'.[26] This sentence captures the great divide between the pre-modern and modern eras, which is why Don Quixote, though an unlikely hero, is quintessentially a modern one. Indeed, Cervantes can be called the author of the modern age. In his seminal novel humanism is uppermost: men fight for themselves, not for God, and for a universal principle, not for their kinsmen.

The difference is critical. Kinship is something that is innate: we are born into it. It is only when we choose for whom we fight that we transform kinship into affinity. In German affinity is not just qualified kinship as it is in English: it is the opposite of it. It is freely chosen. And with affinity is born the modern sensibility – we are now free agents of ourselves. Like the Don, we can choose for which of humanity's causes we wish to fight.

With the coming of the modern age the warrior meme is forced to adapt to a new environment. Excellence gives way to merit. And merit is grounded in a quite different view of human nature: *perfectibility*. Unlike the ancients, the moderns aspired to free man from any natural constraints. The true warrior does not realize his natural (God-given) abilities but rises above them. He is determined to escape from the concept of nature, or limit. He uses reason and ingenuity (technology) to go beyond physical limits, the strength with which he is born. He cultivates a concept of duty and thus escapes his baser nature, the pursuit of war for its own sake. Thus merit becomes the power of disinterested action, acts we do for others, not for ourselves. With the idea of the disinterested hero comes the idea of duty for the realization of the common good.

W. H. Auden expressed it better than most in an essay written in the 1930s. The modern warrior is far removed from the pre-modern because he is subject to a universal law: justice. We always ask of him, in what cause did he die? For a modern warrior to be honoured there must be a moral or historical significance to the conflict, and therefore meaning for us in his success or failure. And the pathos of his death too is different. In Hector the pathos is that the noblest warrior dies for a cause we know is lost. As Auden acutely observed, the Homeric hero lives in a perpetual present, compared with the modern hero who dies for the future. The world of Homer never transcends the immediate moment: one wins, loses or dies. Joy and suffering are of immediate moment but they have no meaning beyond it.

> It is a tragic world but a world without guilt for its tragic flaw is not a flaw in human nature, still less a flaw in individual character, but a flaw in the nature of existence.[27]

The Homeric hero or medieval soldier, like Roland in the *roman de geste* of that name, is one-dimensional. He is born brave and courageous. His area of free choice and thus responsibility for his actions is fairly circumscribed. The modern hero, by contrast, is brave because of a subjective choice: he feels fear and overcomes it, and in overcoming it becomes heroic.

Now, if the warrior puts himself in the service of others and permits himself to be judged by others, he must find the wellspring of duty in himself – he must 'hear' duty as an inner command. He must have a vocation. Agnes Heller puts it thus:

> To be pulled by one's own destiny means to be destined. But this is not destined from the outside but rather destined from the inside by one's own choice.[28]

How unlike Aeneas. You choose to be a warrior and become what you already are (therein lies human growth). One chooses to be good at something and therefore better (not only as good as everyone else). One constantly wants to *improve* and thus outdo, not merely emulate, Achilles. And to be true to oneself one must have self-knowledge as well as an inner life. To lie to yourself is to become a failure, and in failing your profession you fail as a man. Thus Shakespeare's great contemporary Montaigne wrote:

> We must go to war as a duty. The reward we should expect is one which cannot fail any noble action however obscure it may be; we should not even think of virtue but of the satisfaction which a well governed conscience derives from acting well. We must be valiant for our own sakes. . . . Our soul must act her part not when on parade but at home within us when no eyes but our own can penetrate.[29]

His conclusion? We must fight not to immortalize ourselves but in order to be *true to ourselves*.

What explains the change is the birth of a *modern* consciousness. J. H. van den Berg, a Dutch psychiatrist, dates the birth of the inner self to 1520 when Luther discovered 'the inner man distinguished from the physical world'. Harold Bloom, the noted Shakespearean commentator, prefers to date the birth of the inner self to *Hamlet*. Indeed, Bloom claims that he can find in pre-modern literature no character willing to make a free artist of himself, whether Achilles or Aeneas. For neither man is changed by overhearing what he may have said and on that basis turns himself about.[30]

This is what Bloom means when he credits Shakespeare with 'inventing the human', or our humanity, our modernity. In making that claim he is indebted to Hegel's posthumous lectures, *The Philosophy of Fine Art*. For it was Hegel who distinguished Shakespeare's characters from those of Sophocles, who cannot change through self-knowledge because their tragedy (or individuality) is part of a higher ethical power. Shakespeare, by contrast, confers on his characters both intelligence and imagination. This is the importance of the Shakespearean soliloquy. Shakespeare's characters grow, develop and often change in the course of the play. They have self-consciousness. They are men who think of themselves as the source of their own acts. They do not receive their laws or norms either from the nature of things (as Aristotle claimed) or from God (the belief of the Middle Ages) but establish them themselves on the basis of their own reason and will. Man is the subject now because the dignity of the self does not lie in the objective situation assigned to it in the cosmic order. This marks a decisive break with the ancient world as well as with the Middle Ages.

We find some of the most telling evidence of this change in consciousness in one of Shakespeare's later plays, *Troilus and Cressida*, when the Greek generals are discussing the military situation before Troy. Agamemnon takes issue with his generals for their depression. True, the Greeks have met with ill luck, but the gods can be blamed for their misfortune. Nobility in man is found not in success but in the ability to cope with failure. All human beings are alike in good fortune. It is in adversity that their greatness shines through.

> But in the wind and tempest of her frown
> Distinction, with a broad and powerful fan,
> Puffing at all, winnows the light away;
> And what hath mass or matter, by itself
> Lies rich in virtue and unmingled. (1.3.26–30)

This is a philosophy which, while admirable in Homer, is hardly modern: it is really the philosophy of the Cross, or the mystic grandeur of Greek tragedy, in its passive acceptance of fate. And it is a view that Ulysses rightly rejects. In the play Ulysses is a very modern general who chides Agamemnon for refusing to see that the Greeks owe their failure to their own indiscipline. Indeed, in the resentful Achilles ('mocking the designs of the leaders and breaking scurral jests') he sees a soldier who belies his claim to be a true warrior by allowing his passion to overrule his intellect. Skulking in his tent, he is conceived as a man of great strength but weak intellect.

In his wilful egoism Achilles is an absurd figure whose self-indulgent behaviour belies the application of reason, which Ulysses values above everything else:

> Forestall prescience and esteem no act
> But that of hand: the still and mental parts
> That do contrive how many hands shall strike
> When fitness calls them on, and know by measure
> Of their observant toil the enemies' weight –
> Why this hath not finger's dignity:
> They call this bed-work, mappery, closet-war
> So that the ram that batters down the wall
> For the great swing and rudeness of his poise
> They place before his hand that made the engine
> Of those that with the fineness of their souls
> By reason guide his execution. (1.3.199–210)

Merit requires the warrior not to be, but to *become* what he is. And the chief way to do that was through education. So we see in the early modern era the birth of war as an art, and the growth of military reforms beginning with the infantry reforms *all'antica*, 'after the antique', promoted by Bartolomeo d'Alviano (1513) for the armies of Venice. The first important book on the subject, *The Art of War* (1522), was by another Italian, Niccolò Machiavelli, the only work of his to be published in his lifetime.[31]

In time a new ethic of service to the state replaced the more individualistic ethic of chivalry. Not that chivalry itself disappeared immediately. Chivalrous literature was never more prolific than in the sixteenth century, but it ceased to stress the spiritual goal of knightly contest. We still find chivalrous gestures – the challenge to single combat, release without ransom of a particularly gallant foe, and efforts to save women from the horrors of a siege – but most of these episodes are best seen as codes of freemasonry. The idealization of chivalry may still have been pervasive, but in the course of time chivalry evolved into cavalry as the mounted arm became more a separate arm of service and less a military expression of the ruling class. And the military nobility became an officer corps in a hierarchy in which, for the commander in chief, battlefield example mattered much less than generalship. Bravery counted for much, but skill at arms for much more. 'Warfare was now an intellectual problem not an athletic exercise', writes Thomas Arnold, and that pretty much sums up Ulysses's dismissal of Achilles in Shakespeare's play.[32]

Steadfastness under fire became the essence of courage in modern infantry warfare. If officers stepped aside, then the ordinary rank and file would disintegrate and the company and battalion with it. Vantome, whose memoirs tell us much about the military culture of the late sixteenth century, noted that for an officer, 'glory both at court and among women consists of blows received and not given'.[33] Such developments helped to establish war as a serious branch of knowledge and to establish the warrior as a professional, a man who merits the respect of others.

Warrior meme 3: Authenticity

Clausewitz, for one, was keen not to reduce war to blows received and not given, especially when those blows took the form of anonymous firepower. Indeed, he insisted on the importance of such subjective forces as life, character, spirit and genius. In a letter to Fichte (1809) he criticized 'the tendency particularly in the eighteenth century to turn the whole into an artificial machine in which psychology is subordinated to mechanical forces that operate only on the surface'.[34] He wanted to stress the human element as the 'essence' of war in the face of its growing inhumanity, or disenchantment.

By the twentieth century war had become even more disenchanting for the warrior as it began to be defined in terms not of the men who fought it but of the machines that waged it. Soon machines, not men, came to dominate the human imagination.

From an artist's point of view, the key date in the disenchantment of war is 1912, when the painter Fernand Léger found himself accompanying two friends, Marcel Duchamp and Constantin Brancusi, to the Aeronautic Salon at the Grand Palais in Paris.

Duchamp challenged Léger with the claim that 'painting is finished. Who could do better than the propeller? Tell me, could *you* make that?'[35] Brancusi's reaction was to incorporate the dancing curvatures of the propeller within his own sculptures. Léger's response was to celebrate his comrades in the field two years later as titanic machines who consist of the same materials as their guns.

In his pictures war and industry have indeed become the same. The most often repeated images in his work are those of cylinders, pistons and rods, which belong to the battlefield as much as they do to the factory. Léger fought in the French Engineers Corps, so he had direct contact with the machinery of war. He particularly enthused over the 75 mm gun, in the same way Duchamp two years before had enthused over the propeller. He is quoted as saying:

> I was dazzled by the breech of the 75 mm gun which was standing uncovered in the sunlight. The magic of white light on metal. This was enough to make me forget the abstract art of 1912–13.[36]

Invention from then on began to be identified as the primary mode of creation in the modern world. Technology was shown as an active poiesis. Art could no longer rival the *techne* of the engineer. In the words of George Steiner, art 'could no longer think itself through' – it could no longer grasp the human dimension of war.[37] All it could reveal was its inhumanity.

Thus we find no portrait of a soldier to match Leonardo's drawing of a *condottiere*, or no painting that expresses the physical excitement of war to compare with Utrillo's *The Battle of San Romano*. Instead, we have Ginno Severini's *The Armoured Train*, whose focus is not man but the energy of war, and C. R. W. Nevinson's *La Mitrailleuse*, which is as much an evocation of the machine gun as of men in battle. As Nevinson later wrote, 'To me the soldier was going to be dominated by the machine. . . . I was the first man to express this feeling on canvas.'[38] The claim was confirmed in the pictures he painted after being invalided back home, such as *La Mitrailleuse* and *The Motor Ambulance Driver*.

If this was the reality that war had become, the only strategy left to the warrior was a very late modern one: to authenticate his existence on the battlefield by digging deeper into his own subconscious, in realms beyond the machine and the ethos of industrialized warfare. The warrior meme adapted itself (as all memes do) to a new environment in order to survive.

What we are discussing is another change in modern consciousness, a feature of late modernity. By the time Freud had formulated the chief principles of psychoanalysis in the late nineteenth century, both the general notion that there are aspects of the self which are not conscious and the specific term 'unconscious' were already well established. Freud himself was following a cultural trend, not starting one. The unconscious was familiar to Schopenhauer, Nietzsche and others: the idea that humans are frequently unaware of their own motives and driven by impulses and needs they do not always acknowledge. This is the understanding of the self as a complex, by no means entirely rational, entity in which impulses are everything. Both the idea of the unconscious and the idea of a collective unconscious – such as a national destiny, or a will to power, or a people, or history – can be found in the works of writers as distinctive as Ernst Junger.[39]

The best example of that dialectic can be found in one of the poems that forms part of W. B. Yeats's collection *The Wild Swans of Coole* (1919). 'An Irish airman foresees his death' describes the freely chosen death of an Irish airman, Major Robert Gregory, who lost his life on the Italian Front in the closing months of World War I.

> I know that I shall meet my fate
> Somewhere among the clouds above
> Those that I fight I do not hate
> Those that I guard I do not love;
> My country is Kiltartan Cross,
> My countrymen Kiltartans' poor
> No likely end could bring them loss
> Or leave them happier than before.

> Nor law nor duty bade me fight,
> Nor public men, nor cheering crowds,
> A *lonely impulse of delight*
> Drove to this tumult in the clouds;
> I balanced all, brought all to mind
> The years to come seemed waste of breath,
> A waste of breath the years behind
> In balance with this life, this death.

'A lonely impulse of delight' – it is a striking phrase, one that sums up the attempt to assert the warrior ethos in the face of the reality war had become by 1918. For the war in the air, of course, dangerous though it was, was also removed from the reality on the ground, and the disenchantment it brought. Equally striking is Yeats's reference to the way in which an empty life can be balanced by an 'authentic' death.

To overcome alienation re-engagement in life is required, and for the true warrior this took the form of battle. What was an 'authentic' death other than one that conformed to the belief that death threw life into relief, war as a *transcending* moment, even if a short-lived one. A literary example can be found in Thomas Mann's greatest novel *The Magic Mountain* (1924), when the hero Hans Castorp is seen for the last time running into battle, bayonet in hand, only to be blown off his feet by a howling 'hellhound . . . a huge explosive shell, a disgusting sugarloaf from the infernal regions . . . the product of a perverted science'.[40] For what was especially alienating for the modern warrior was the marriage of science and war that found its apotheosis in World War I. Released from the sanatorium in which he has been living for years, Castorp dies contentedly in the trenches. He doesn't seek his death, but he doesn't try to avoid it either. Indeed, he dies content in the knowledge that through war he has understood life for the first time.

As for collective destiny, Junger expressed it well. 'One great destiny carries us on the same wave. For once we acted together as a single organism facing the hostile external world, men . . . bound together by a higher goal'.[41] Castorp too leaves the sanatorium to go to the Front – at which point Mann's novel ends because his fate now belongs to another story, another goal: that of world history. His end is similar to that of Antoine in Roger Martin Du Gard's masterpiece, *Les Thibaut*. Gassed and crippled, his death is still authentic. As Camus added in a review of the book, he 'vanishes into the very stuff of history, of which men's hopes are made and whose roots are human misfortune'.[42]

As the world entered the nuclear era the authenticity meme came under intense challenge. One of the last writers to embrace it was Julius Evola, a man who is regarded by many as the godfather of contemporary Italian neo-fascism. Evola's book *Men Among the Ruins* is a frontal assault on the prevailing materialism of the time; that it was written as recently as the early 1970s is a remarkable example of how tenaciously the political right held on to the concept of authenticity even in the nuclear era.

In his work Evola hyped up the virtue of 'militarism and military values'. He extolled the 'higher right of a warrior view of life'. For him, as for Hegel, the warrior embodied an ideal type, a heroic personality. Evola did indeed acknowledge that modern war had become soulless, a war of the machine against man, but he believed its soullessness could be overcome by authenticity, or what he called *karma*. Like Junger, he believed a warrior had to come to terms with the machine age. What was needed was a 'cold, lucid and complex heroism' in which the old romantic, instinctive, patriotic aspect would be replaced by 'a sacrificial disposition: man's capability to face and even to love the most destructive situation through the possibilities they afford . . . possibilities, [which] in their elemental character, offer him the chance to grasp what may be called the "absolute person"'.[43]

In the end, of course, the attempt to 'authenticate' the warrior was as self-defeating in the 1970s as it had been in World War II. An earlier generation of existentialists like Heidegger, Sartre and Camus had insisted we accept that alienation is an objective condition of life. Part of what it is to be 'authentic' is to grasp, accept and, perhaps, even affirm this fact. For the existentialist the task was not to overcome the feelings of alienation by seeing the world as a home and oneself as 'homeless', but rather to have the courage to live in the clear consciousness that it was not a home, and never would be. Others like Evola sought to end the alienation of the warrior in the face of his humbling on the industrialized battlefield by adopting strategies which denied the materialization of war and emphasized the transcendent: the 'will to power' by which humanity could transcend material conditions and thus preserve war for 'life'. But in dealing with abstractions such as 'the destiny of a people', or 'Being', or 'History', or the 'nation state', they all tended to humble man to a greater extent than ever. In attempting to re-enchant war, they merely added to its general disenchantment, and they increased the alienation of the warrior by prompting him to invite dangers and welcome threats, if not reject a life considered 'inauthentic' or unworthy of living.

The Biotechnological Warrior

> Who than can live among this gentle
> obsolescent breed of heroes and not weep
> Unicorns almost?
>
> Keith Douglas, 1943

I have painted a large canvas, one that is not only wide and various, extending into history and literature, but also crowded with the names of heroes, both fictional and real. I have done so to illustrate how tenaciously the warrior meme has persisted – until now. For with the evolution of war into a new age – our own, the age in which the future is being forged – genetics is beginning to take over. The picture, of course, is more complex than that. The information and biotechnological revolutions are interlocking.

To illustrate this let me cite two passages from James Blinn's influential novel of the first Gulf War (1990–1), a vivid portrait of a virtual warrior who finds himself taking part in history's first 'virtual war'. Blinn's dazzling book is the most evocative work on war since Joseph Heller's *Catch 22*, and while Heller captured what was essentially a World War II soldier's response to what war was in the process of becoming, Blinn's is a forward-looking satire on what war has now become.

Set on a gigantic computerized aircraft carrier conveying a state of the art arsenal bristling with SMART weaponry to the Persian Gulf in history's first truly TV war, the novel evokes what war means to a twenty-first-century American warrior such as Blinn, who served in the US Navy for nine years. Asked what he does for a living, the hero replies:

> She asks what I do and I'm tempted to give her a dose of some acoustic techno-wizardry airborne and anti-submarine warfare jargon. Initiate a little battle of the jargonauts. Lay on some acronyms and abbreviations: ASW, FASOTRAGRU-PAC, ECS, MK-82, ADP, INCOS, SENSO, TACCO, COMNAVIRPAC, ECP, NATOPS, ESM, MAD, SAP, ACLS, AN/ALR 47, ASWWINGPAC. Or just gab along in navpubspeak: *The mission of the Sierra-3 Viking fixed wing, carrier based, all-weather, tactical anti-submarine warfare aircraft is to utilize its suite of active and passive computer-assisted detection sensors to localize, track and terminally engage surface and sub-surface hostile contacts in electromagnetically charged combat environments . . .*
>
> . . . Instead I just say: 'I'm in the navy.'[44]

Such is the authentic vernacular of the twenty-first-century warrior. If the language is technical, this befits a soldier or pilot who occupies the same virtual space as his own weapons system.

Blinn's hero goes on to describe a world which he shares with a computer-machine, a world in which the traditional biological instruments such as the human gaze are now largely machine mediated. In a memorable scene he records how he saw a dead man for the first time with his own eyes, not through the usual medium of technology.

> I saw it with my own eyes. I saw it with my EYES.
>
> Therefore it's real. Isn't that how it goes? If you see it with your own eyes, in the same time and space as you're in, then it's real. If/then. Causality. That's it, right? It wasn't relayed or bounced or fibre-optic transmitted, modulated or written or echoed, encoded, encrypted, or ciphered, projected through celluloid, optically etched on silver halide, simulated by ordering the polarities of magnetized ferrous atoms, facsimiled, thermal, laser or holographically imaged, analog or digitally processed, scan-converted, manipulated, synthesized, distorted, Animatronicized, equalized, morphed, tweaked, computer-enhanced, duped, dubbed, multiplexed or multitracked, photo-copied, mimeographed or inkjet printed, colorized, SurroundSounded, Dolby-ized, virtually-realized, or electron-energized on the back of a CRT . . .
>
> . . . No filters, no intermediary, no question of interpretation.
>
> Authentic, three dimensional. Hard-wired. In my face. The stink of death.[45]

In both passages (there are many other telling observations in the book) Blinn offers us a graphic picture of what war has become, and with it the warrior. In the next chapter I shall discuss the technological matrix of man and machine as the two continue to co-evolve equally. Here I am concerned only with how it has redefined the *existential* dimension of war.

For the warrior meme has readapted to its environment again as human and machine are increasingly assimilated. What we are witnessing is the interaction of man and machine, as both respond to the demands of a continuous stream of information. It is feedback which allows the system to function. In this cybernetic world there is no question of finding one's 'authentic' being because there is no 'essence' of humanity, any more than an 'essence' of war with which we can put ourselves in touch. Human nature and intelligence are dynamic and interactive. They are *performative* because both organic (human) life and machine life work as one (cybernetically). Both exist by responding to external stimuli, but they respond very differently to how they did in the past.[46]

Cybernetics is the science of a self-regulating system (one that corrects its own errors). A machine that is self-regulating and self-correcting is called a 'servo-mechanism'. Feedback (the use of information provided to determine the effectiveness of past and future actions) is essential because the system guides its own behaviour, using the results of its own past and present performance to determine its future performance. In this respect, it differs from stimulus response (SR), which determined the performance of soldiers in the past. In SR when a stimulus is presented, an organism (soldier) responds. A new stimulus presents itself and the organism responds in the same way at the same time. This was the basis of the very different memes of 'excellence' and 'merit', and of 'authenticity' too. Excellence was predictable behaviour, and professionalism was based on it, just as authenticity was a disciplined response to a pre-programmed stimulus. In both cases there was no inner processing, just an automatic response.

We can date this change, as we have seen, to the work of Norbert Wiener, the father of cybernetics, who began work in 1941 on the firing mechanisms for anti-aircraft guns. Recognizing that if they aimed directly at a fast-moving plane they would miss every time, he realized that the only way to score a hit was to ensure that the guns were self-adjusting. Only by estimating where the targets were and how fast they were moving could they be targeted successfully. Information about their speed and direction had to be fed back into the aiming mechanism.

As Wiener wrote, 'To live effectively is to live with adequate information. Thus, communication and control belong to the essence of man's inner life even as they belong to life and society'.[47] The same can be said of the twenty-first-century warrior.

> A modern fighter pilot is a technological breeding ground for a new kind of site-specific 'self'. This new subject position and subjectivity that operates under the name of 'fighter pilot' is however not solely the product of its own unique institutional opportunities, specialized domains of knowledge and technology. It is also the product of a

> psychasthenic logic. For the fighter pilot's model body and its technologically articulated consciousness exist in their purist state at the site (the fighter plane's cockpit) of a fundamental disturbance of perception, that of a schizophrenic of self and place, the result of an organism's almost perfect assimilation to its surrounding space.[48]

The man–machine interface, the new environment or computer space which machine and man inhabit together, is not an extension of the body (like a tool) but a total environment. It is the context for a new corporeal reality, an entirely new world in which war is conducted, a world into which we are sensorially (not only physically) incorporated and assimilated.

For it is important to recognize that war is not becoming disembodied. This is especially true (surprising though it may seem at first glance) of virtual reality, for which much of the early groundwork was done in the 1960s by the US Air Force (USAF) in its design of flight simulators. Computer graphics today have such a high degree of realism that their sharp images evoke the term virtual reality (VR). But VR does not merely 'replicate' reality. It sucks you in; it immerses you. The computer generates much of the same sensory input that a jet pilot could experience in an actual cockpit. In accessing the virtual world, the pilot is interfacing with the computer.

Looked at in this light, the re-enchantment of war for the warrior is to be found not in the computerization of the battle space or even the 'virtuality' of war, which might indeed render the warrior as obsolete as the 'unicorns' Keith Douglas evoked in 1943. For whilst some artificial intelligence (AI) specialists argue that it is only a matter of time before computer programs are able to simulate human experience such as pain and emotion (and presumably learn from them), computers are still far too dependent on adherence to formal procedures to develop such human traits as intentionality, imagination or understanding.

Intelligence is nevertheless ontological. For it is still embedded consciousness that relies on biological factors such as neurological synapses and bioelectrical currents. Consciousness is an embedded phenomenon – the mind learns within a material environment.[49] Computers are hard logic devices; only human beings have intuition and emotions that derive from embodiment. In war, as in life, spontaneity still prevails over programming. War is still an existential experience, though it is a different one from the world of Homer and Montaigne, or more recently Junger. For we are beginning to privilege the biological over the cultural, our genes over our memes.

Let me go back to what I said about the existential dimension of war. The warrior, Hegel tells us, is a special kind of man, one culturally programmed like the rest of us to want the recognition of his peers, but one whose desire can only be fulfilled through war. For the warrior is a specific human type and until that type disappears it would be foolish to expect war to ever come to an end. In his willingness to put his life on the line for a warlord, feudal master or king, or more recently a people, the warrior defines himself as a man. Courage is his currency as a human being. It is as much a cultural construct, of course, as it is a biological one, for it is in our nature to try to survive. Warriors are not warrior ants programmed to die for a colony which happens to be their best chance of survival, given that they share 95 per cent of their genes with their cousins. As human beings we only survive if we pass on our genes to our children. Yet remarkably, warriors are willing to risk everything at their most sexually active age. Their willingness to sacrifice their lives is not biological at all, it is social. It is at odds with their animal nature.

This interpretation of Hegel owes much to Alexandre Kojève, a Russian émigré author of the 1930s who gave Hegel's writing an enduring twentieth-century gloss. Man, he adds, is an animal, but through his willingness to risk his life for a non-material reward such as social recognition or social status, and above all the esteem of others, he rises above his animality, as we all aspire to. Interestingly, the two examples Kojève cites in his reading of Hegel are a medal and an enemy flag. A soldier fights to secure both not for their practical value – they have none – but because they are *desired* by others. Warriors live in the recognition of others, usually their fellow citizens, more often their peers, and sometimes even their enemies, at least those that they respect as worthy adversaries.

It is for that reason, however, that the warrior meme is in trouble. For our culture puts enormous emphasis on the avoidance of pain. Those who inflict pain on others are despised. As Nietzsche correctly tells us, civil society's main ethos is a contempt for cruelty. And soldiers are expected to avoid pain too, consistent of course with fulfilling their mission. Kojève's great admirer, Francis Fukuyama, asserts in a wistful passage towards the conclusion of *The End of History and the Last Man* that

> In our world there are still people who run around risking their lives in bloody battles over a name or a flag or a piece of clothing but they tend to belong to gangs with names like the Bloods and the Cripps and make their living dealing drugs.[50]

For Fukuyama, as for Hegel, a man is most human when he risks his life, when he allows his human desire to prevail over his natural instincts. In showing his contempt for life at any price, he puts human dignity first and in so doing becomes a true moral agent. In affirming his agency he reaffirms his humanity.

But in a society in which soldiers are no longer encouraged to discover their self-worth through the esteem of other men – or perhaps more correctly, are not allowed to – western societies may only be able to engage in war not by emphasizing the cultural (the meme) over the biological (the gene) but by *emphasizing the biological at the expense of the cultural* – by re-engineering warriors through biotechnological means.

In civil society that trend is already in evidence. *In vitro* fertilization may represent only 1 per cent of births in the United States, and embryo selection a proportion that is not much larger. Cloning and genetic modification clearly lie in the future, but the future can be glimpsed. We are escaping a humanity defined by natural selection. We are on the cusp of 'overcoming' our humanity, to use a word favoured by Nietzsche. In determining our own evolutionary future, we stand on the eve of a 'post-human' age.

By drawing reproduction into a highly selective social process that is far more successful at spreading successful genes than sexual competition, we truly are embarking on a new voyage. Within the next 50 years we will be able to modify ourselves, to design our own babies, and possibly to create better soldiers. The technological powers we used in the past to change the natural environment can now be directed at changing ourselves, by modifying not so much human nature as the behaviour of specific human types, the warrior included. This is made possible by breakthroughs in the matrix arrays called DNA chips, which may soon be able to read 60,000 genes at a time; the manufacture of artificial chromosomes, which can now be divided as stably as their naturally occurring cousins; and advances in bio-informatics, the use of computer-driven methodologies to decipher the human genome. The latter include old-fashioned steroids as well as new drugs that will enable us to enhance our physical capabilities.[51]

The enhancement of athletic abilities by drugs has been with us in sport for a long time. Take the banned hormone erythropoietin, which by raising the oxygen-carrying capacity of red blood cells can boost endurance by 10–15 per cent. Metabolic and physiological enhancers are now a central part of sport. The only question is their detection, and their side effects. The social pressure to re-engineer the athlete – for the athlete to win and to win more spectacularly than in the past – has generated the relentless use of pharmacology in sport. Likewise, the equally insistent need to win is likely to accelerate the use of pharmacology in war, as we will see in the next chapter.

And then there is genetic modification. Already, we can modify a trait in a directed fashion by altering or selecting particular gene variants. Changing a single gene in an animal is now routine. The research has been spurred on by scientists' claims that they can decipher the relationship between genes and behaviour such as criminality, alcoholism and drug addiction. The trick is to identify a combination of gene variants common to many people with similar endowments (such as athletic prowess), and then to manipulate the human germline. Is there a gene for aggression? Is there a warrior gene as well?[52]

With so much genetic information available on every human being, from simple single-gene disorders to complex polygenic moods and behaviour traits, it is becoming attractive for employers to use genetic data to select prospective employees. As early as the 1970s the discovery of the sickle-cell anaemia trait prompted the armed forces to use genetic screening for the first time. Carriers of the recessive gene – most of them African-Americans – were denied entrance to the USAF Academy for fear that they might suffer the sickling of their red blood cells in a reduced-oxygen environment.[53] The US military purportedly went further in 1992 when it launched an ambitious programme to collect several million DNA samples from its personnel. The exercise was to enable the accurate identification of men and women lost in combat. But in the legal battle that followed the refusal of two marines to give blood under the Fourth Amendment right to privacy, the fear was expressed that the same genetic samples could be used for biomedical research to identify the best military genes, or to weed out soldiers with the worst: those most susceptible to fear.[54] If it is possible to isolate genetic traits, then it should also be possible to enhance personality traits such as risk taking that would be required by Special Forces, and to produce above-average levels of emotional stability for pilots in the virtual spaces they occupy with computers.

Will this result eventually in the emergence of a warrior elite, a caste genetically distinct from other non-combat units? Traditionally the military has often seen itself as culturally distinct in its attachment to a value system that still honours courage, heroism and indeed honour itself. But in a biotechnological age, culture may be transcended by biology. With the emergence of genetic screening, why demarcate a warrior by class or by ethnicity? Why not do so on the basis of genotype, on the basis of positive discrimination – isolating certain 'positive genes' – or negative discrimination – screening to detect predispositions to mood and behavioural instability?

Similar questions are posed by cyborg technologies – the technological as opposed to genetic enhancement of the body. In re-engineering themselves, will warriors come to see themselves as members of an exclusive class? The process of 'technologizing' – in which bodies are reassembled so that they can function more optimally (i.e. so that excellence can be enhanced) – is central to the cyberpunk sci-fi of William Gibson. In his imaginary world, cyborgs are creatures whose identities are no longer determined by social criteria (such as class) or even by ethnicity but by *technicity* (the new architecture of the body).[55]

In Gibson's world cryogenic processes and enhanced digitalized senses redefine identity, just as cyberspace produces its own virtual communities. In 'Johnny Mnemonic' one of the principal characters has electronically upgraded vision and prostheticized fingers that house a set of razor-sharp, double-edged scalpels, myoelectrically wired into her enhanced nervous system. She is no longer an individual (born into a social or ethnic group from which she derives her sense of self); she is a customized, functional product of a cyborg culture, and she has little respect for others who are not like her. What Gibson offers us is a vision of a separate caste – a world in which the respect one warrior has traditionally given another is no longer a product of culture but of bioengineering. What his cyborgs admire in each other is technical virtuosity and operational speed, attributes that have been directly integrated into their own prosthetic and genetic architecture. Here 'excellence' has morphed into what David Tomas calls a 'technological edge'.[56] It is an edge which might one day distinguish warriors not only from civilians, but also from all other warriors who have ever lived. Not for them the Achillean meme.

Conclusion

It is a challenging future indeed, one that many soldiers themselves will find disturbing, for two reasons.

First, war – as we have seen – has a history, and the warrior is a vital part of its *historicity*. War has shown historical development from the Bronze Age verities of the *Iliad*; the piety of Aeneas; the spirit of the crusading warrior; the self-conscious warrior depicted by Shakespeare and Montaigne; and, most recently, the alienated public servant. And its historicity is not at an end – yet. But that history has always involved a human element because of the true warrior's ability to import vitality to so much he touched. His exploits, however heroic or even self-sacrificing, were also received with suspicion because of his own self-absorption and capriciousness. If the object of genetics is to produce a warrior whose behaviour can be totally predicted, then the human element will be called into question.

Secondly, what distinguishes the warrior from every other person who uses violence is the ethical world he inhabits. The warrior typology, of course, includes 'warrior-killers' who live outside that bounded world; whose conscience is so narrowly circumscribed that they are entirely self-referential. But the majority of warriors throughout the ages have emulated those deemed to represent a Nietzschean 'higher type'. Hence the importance of the warrior meme as it fed into an aristocratic code of conduct. The danger in geneticizing the warrior as a caste lies in challenging all models, stereotypes and previous norms at the risk of transforming the warrior himself (or herself), and by extension his or her profession, into a purely utilitarian one with little or no *normative* appeal.

Yet the picture is not quite so clear cut. It is true that in the research laboratories of the United States scientists seem intent on re-enchanting war by reducing to a minimum its traditional existential appeal. War already no longer affords that satisfaction of spiritual needs which the late modern era looked to it to find: the pursuit of national destiny, or national rejuvenation, or an aristocratic will to power. In the West it no longer even fulfils the spiritual needs of a community. But that does not mean it no longer fulfils a purpose or, more to the point, that it cannot be 're-enchanted'. In the next chapter I hope to show that it would be mistaken to conclude that as a species war will have nothing further to say about our future. Much will depend on whether the next phase of war – the post-human – is likely to see a taking further of our humanity, rather than an 'overcoming' of it.

3

Towards Post-human Warfare

Cybernetics and the World War II Poets

The very phrase 'war poet', wrote Osbert Sitwell (a war poet of sorts himself), indicated a strange twentieth-century phenomenon, the attempt to combine two incompatibles, poetry and war. For there had been few war poets in the immediate past, during the Napoleonic, Crimean or Boer wars. But then,

> War had suddenly become transformed by the effort of the scientist and mechanitian [*sic*] into something so infernal, so inhuman that it was recognized that only their natural enemy the poet could pierce through the armour of horror . . . to the pity of the human core.[1]

Well, there may not have been war poets as such before World War I, but war has always been a theme of poetry. Writing on the eve of World War II, Simone Weil celebrated the *Iliad* as 'the purist mirror of our collective experience',[2] and what the World War I caesura marked was not the sudden eruption of poetry into war, but the failure of so many soldiers who were poets themselves to find its 'human core'.

Indeed, what is so fascinating about the poets of World War II (who have far more to tell us about the future face of conflict than their World War I predecessors) is how difficult it was for many to find the war's human dimension, even though they found the pity soon enough. In World War I poems the soldiers have individual names – Yeats's Maj. Robert Gregory, Graves's Sgt-Maj. Money and e. e. cummings's Olaf, 'glad and big whose warmest heart recoiled at war'. Even the unnamed sergeant in Ivor Gurney's 'The Silent One' who 'died on the wire' is given a distinctive voice, the British accent he spoke in life. But in World War II there is only 'The Ball Turret Gunner' or 'The Wingman'.[3] What the poets after 1940 found instead was a 'post-human element' at the core of their experience of war so remote from traditional poetic concerns that they were able to capture the experience without fully understanding it.

By the end of the war, many poets had begun to suspect that technology would eventually subvert war as a human experience. The 'real' war was now not on the battlefield but off it, or what we would now call the 'battle space'. War, wrote the gunner poet Barry Amiel, was reduced to 'a matter of mathematics' with the chances of life or death varying on the curves of a parabola, the scientific soundness of a choice.[4]

One of the most talented poets of all, R. N. Currey, a South African serving as an artillery officer with the Eighth Army, noted the restraint in language and emotional range of his own generation of poets when compared to those like Siegfried Sassoon and Wilfred Owen. This largely explains why the World War I poets were much more popular with the public at the time and have remained so ever since. To be sure, poets like Roger Keyes, Alun Lewis and Keith Douglas were favoured by the public after the war, but the reason is in itself illuminating. For whether like Douglas they served in tank units, or like Lewis with the infantry, whether they served in frontline units whose combat experience matched that of the Great War poets, all three were to be found 'in every position of dramatic personal responsibility and personal danger'. Yet Currey also suspected that the readers of the future would find their graphic accounts of war increasingly remote and emotionally distant from what war was in the process of becoming.[5]

For World War II did indeed mark a major disjunction between the soldier and the technological dynamic of modern conflict. Currey himself admitted that his own imagination was fired by his technical proficiency in operating his radar system, as well as the beauty he found in the mechanism itself: 'the spots of light on a glass screen', the phosphorescent image to which the enemy had been reduced.[6] In the first of two sonnets, 'Unseen Fire', he paints a vivid picture of the new warrior:

> This is a damned inhuman sort of war.
> I have been fighting in a dressing-gown
> Most of the night; I cannot see the guns
> The sweating gun-detachments or the planes.
>
> I sweat down here before a symbol, thrown
> Upon a screen, sift facts, initiate
> Swift calculations and swift orders; wait
> For the precise split-second to order fire.
>
> We chant our ritual words; beyond the phones
> A ghost repeats the orders to the guns;
> One Fire . . . Two Fire . . . ghosts answer; the guns roar
> Abruptly; and an aircraft waging war
> Inhumanly from the nearly five miles height
> Meets our bouquet of death – and turns sharp right.[7]

Such images, complained Philip Gascoigne, seemed to characterize the 'dream perspectives' of war in the third dimension. There was little intersubjectivity here with the enemy. 'Destroying the enemy by radio, we never see what we do', complained another poet, Alan Ross.[8] Instead, the main relationship was an increasingly close one between man and machine.

This was clear in the only really romantic battle of the war – the Battle of Britain. For however eulogized the pilots were at the time, the real hero was the Spitfire, a machine that was seen as an animate combatant, a plane that was of course a hybrid: an interface between man and machine. In the popular imagination the planes were anthropomorphized.[9] Indeed, the Spitfire so captured the British imagination – the public set up Spitfire funds to buy more – that it became the icon of the battle, the equivalent of Achilles's armour and Arthur's sword. Except that in this war of machines, it was the machines that were mortal.

Even on the ground the soldier began to feel he was becoming increasingly disengaged from conflict, or dispossessed of war as an existential experience. Some fighting men began to think of themselves as merely an extension of the machine, an adjunct for some limitation the machine possessed on account of its incomplete development.[10] In a poem entitled 'Soldiers', one poet records how he felt increasingly estranged from the 'man beneath this fighting skin'. Indeed, the skin had become compacted to his weapon, by 'the same dream enmeshed / and the one lust famished'.[11]

The political scientist C. Wright Mills coined the term 'cheerful robots' for the type of soldiers now demanded of the times. In the years that followed the war, 'systems analysis', 'social psychology', 'behavioural sociology', 'personnel management' and 'computer-mediated systems' ensured that soldiers became integrated even more into their weapons systems, and they have remained so ever since.

The poets I have quoted were, of course, late modern men. When they thought of war, they first sought to locate it in a traditional ethos that still drew inspiration some 3,000 years later from Homer's classic epic. That's why many were so overwhelmed by their experience of war, in some ways even more so than their Great War predecessors. What depressed some was their suspicion that T. S. Eliot's 'Hollow Men' might be replaced by 'armoured men of metal and of steel', a phrase that appears in a poem by G. S. Fraser: 'men who neither think nor feel / whom we poor poets curse'.[12] And curse them they did, not because war had finally transcended the power of poetry (both epic and lyrical) to capture it. What had happened instead was that war had begun to be transformed in a way that is more apparent to us now, looking back, than it was to the soldiers who fought World War II.

What the World War II poets perceived (though they could not articulate it in technical language) was the new science of cybernetics. In future, war will still be *ontologically* important, but its importance will lie in its cybernetic language. For soldiers are becoming a mixture of machines and organisms. Nature is being modified by technology which, in turn, is becoming assimilated into nature as a functioning component of organic life. The traditional split between man and machine, as well as between man and nature, is fast disappearing. Not only are many of the new technologies calling into question the immutability of the boundaries between the two, but they are also testing the barrier between the 'artificial' and the 'natural'. In challenging the fixity of human nature, our digital and biotechnological age is forcing us to examine all our assumptions about personal identity, including our *uniqueness* as a species.

What is called 'the post-human condition' is a going beyond 'natural selection' to human involvement in our own evolution, what has been called 'participatory evolution'. It also involves an increasing interface between man and machine, or what we call the cyborg condition (familiar to those 10 per cent of Americans who have pacemakers, artificial joints and implanted corneal lenses for computer-assisted sight). Scott Bukatman has called this condition 'terminal identity', writing that it is 'an unmistakeably doubled articulation' that signals the end of traditional concepts of identity even as it points towards the cybernetic loop that may generate a new kind of human subjectivity which makes up the existential dimension in war.[13]

Not that we have thought about that dimension much in recent years. On the atomic battlefield there was no place for a soldier. And in the last few years all the talk has been of Revolutions in Military Affairs (RMAs). Of the 13 RMAs so far discovered by historians, most if not all have been addressed largely in instrumental terms. But most – in one way or another – impacted upon the warrior's view of his own profession. The use of the longbow, followed by the introduction of cannon, destroyed the idea of chivalry and active courage. Courage became more passive in nature; it was valued in a new currency: of blows received and not blows given.

The rise of mechanization on the industrialized battlefields of the late nineteenth and early twentieth centuries locked the warrior into a system in which his performance was increasingly evaluated in industrial terms: productivity and predictability. The warrior became a 'worker'. In our information age he has become an information processor, locked into a cybernetic world. And the biotech revolution promises to transform him again, perhaps in more radical ways still. This is the theme of the present chapter.

Of the many technologies that are changing the warrior's sense of 'self', three are essential to his or her 'post-human' future.

1 *Performative*: The phenomenology of human–machine interaction is changing as computers become more interactive and sophisticated. The task is to make us more machine friendly, if not quite to see war from the machine's point of view.
2 *Behavioural*: We have begun to turn the analytical tools of molecular biology into engineering tools. Most of the technologies today are compensatory (i.e. they compensate for injury or degradation of the body – spare parts, cosmetic surgery). Tomorrow, the human body will be enhanced through the fusion of organic and cybernetic material. The cyborg condition will allow us to enhance our innate abilities, to do much better the things we have always done. One of the things we have always done well, of course, is war.
3 *Normative*: Genetic manipulation and the use of synthetic drugs are rapidly extending the range of human actions beyond what is achievable by natural selection. But both may also influence the way we think about ourselves in relation to others, especially enemies.

In all three cases technology no longer involves an extension of the human body, as has been the case since the first tools and weapons were invented. It is now being incorporated or assimilated into the body at an increasingly fast pace. The technological dynamic is changing in ways that Heidegger could not have imagined even at his death in 1976.

Is Computerizing War the Way Forward?

The network world in which we now live is encouraging us to rethink our biology and the state of our bodies. We have already begun to think of ourselves in terms of machines (though not in the disenchanting way of old, thinking of ourselves as subordinate to them). The computer has begun to change the way we think about our minds, our consciousness and even our selves. Many of us, after all, talk of 'programming' our lives and 'interfacing' with others.

Let me take just one example from an exchange in William Gibson's novel *Neuromancer*:

> 'Can you read my mind, Finn?' . . .
>
> 'Minds aren't *read*. See, you've still got the paradigms print gave you and you're barely print literate. I can *access* your memory, but that's not the same as your mind.'

Gibson offers us two different metaphors of consciousness. The first interlocutor, Case, thinks of consciousness as a book that can be read. His friend Finn insists that consciousness is like a computer program. Gibson's novel spells a computer-mediated end to the old model of human subjectivity which some of us still cling to.[14]

That world once included the world we saw for ourselves. For some time now computers have been doing the seeing for us. This emerges powerfully in Joseph Heller's *Catch 22*: 'It means nothing but you'd be surprised at how rapidly it's caught on. Why, I've got all sorts of people convinced. I think it's important for the bombs to explode close together and make a neat aerial photograph.'[15] General Peckem's search for the ideal bomb pattern captures a reality unique to modern times. For since the introduction of aerial photography at the Battle of the Marne (1914), machines have interpreted reality for us. Our vision is increasingly mediated through technology. It is increasingly synthetic.

In Heller's novel what constitutes success for Peckem and his subordinates is not the number of positions captured, or even the number of enemy soldiers killed, but an image frozen in time, and in this case an objectively meaningless image at that. What the camera captures may not be objectively real – nothing may have been achieved by the bombing, but it is real enough for those who derive their sense of the real from analysing photographic images. It is they who determine success or failure. Satire apart, Heller's novel captures an important reality of war. The computer does most of the seeing for us.

In war synthetic vision has a longer history than we might imagine. It actually dates back to the first military telescope used by Maurice of Nassau in 1609, the man who also reintroduced drill into European warfare. The telescope was synthetic because it required human agency to make sense of what was seen. Indeed, it was not actually an extension of direct vision so much as its displacement. For one has to know what to look for in order to make sense of what one sees. One's vision has to be *informed*. In that sense, the mind and eye are not antithetical. We think because we see, and we see because we think (i.e. we make sense of what we see). The only problem is how to make sense of what we see when the data we collect are increasing exponentially.

One of the first people to appreciate the fact was the poet Goethe. He has been criticized for rejecting Newton's work on optics and claiming that the use of the telescope confused the mind. Stated bluntly, and out of context at that, his objections do indeed sound like obscurantism. But we must remember that Goethe was not only a poet but recognized in his day as a scientist of distinction. What he feared was empiricism run riot. For he was appalled at the thought of being deluged with data which, however accurate, would obscure the structure of things and the regularities of natural phenomena. What he feared most was what today we would call 'information overload'. But he did not anticipate the invention of the computer, which could process, and make sense of, the mass of data coming in.[16]

It is all the more ironic, therefore, that the first computer was conceived during Goethe's own lifetime. Designed by Charles Babbage in various bits and versions in the 1840s, it was finally constructed in plausible form in the closing years of the twentieth century. A working model can be seen in the Science Museum in Kensington, London, a Victorian machine that no Victorian ever set eyes on.

It is a remarkable machine, but to our eyes a very strange one. Babbage's 'Difference Engine' was powered by steam and was intended to compute and print out numbers for such needs as nautical charts. But Babbage imagined a machine that could do more than just compute specific numbers for tables. He envisaged one that could both calculate and store numbers as well. He chose as his model the jacquard loom, which could weave any variety of patterns from information recorded on a set of punched cards (the forerunners of the computer punched cards of the 1950s and 1960s). Babbage called this device the 'Analytical Engine'. Neither the 'Difference' nor the 'Analytical' Engines were ever built in his lifetime. But the first design is brilliantly realized in a novel of the same name by William Gibson and Bruce Sterling, a dystopian vision of a future in which nineteenth-century utilitarianism runs riot.

In the novel a mass of data moves across the country by telegraph on punched cards and spools of ticker tape. In the Central Statistical Office serially connected Analytical Engines process dossiers on every man, woman and child in the kingdom. 'How many gear yards do you spin here', asks one character. 'Yards?' another replies incredulously. 'We measure our gearage in miles here.' What the novel offers its readers is counterfactual history at its best and most frightening, a might-have-been, but one close to the spirit of the mid-Victorian age, to the great dream expressed in an 1840s manual on technical breakthroughs which described the Analytical Engine, even in the absence of a working model, as bringing 'metal close to rationality'.[17]

What made Babbage unique was that he was the first to embody in his design the principle of a general-purpose computational device. For the first time a mathematical rule was successfully embodied in a mechanism whose successful operation required no knowledge of its internal working or the mathematical principle on which it was based. Clearly, it was not the first automatic device to be originated. Textile machines and steam engines had been invented many years before, but these had only replaced physical human activity. Babbage's 'Analytical Engine' would have been the first 'thinking machine' or, put another way, the first incretion of the machine into human psychology.

For its importance also stems from the fact that it was the beginning of the long process by which humans no longer felt they had penetrated the future, but that, as technology advanced, so the future had begun to penetrate them. For what gives the computer its essential role in war is its hypernature – its speed. And it is speed that is now the principal characteristic of warfare.

The problem is that the carbon-based brain does not run on the digital clock cycles of its silicon counterpart. The human brain is not as quick in its reactions as is a computer. On the basis of tests in the 1970s (attaching electrodes to a subject's arms or brain to measure nervo-conduction velocity), psychologists discovered that the speed of electrical impulses through the nervous system is pretty slow for all of us, even the most intelligent: 160 feet per second. Speed, which is becoming central to the texture of our lives, can only be handled by computers.[18]

Computer power is indeed remarkable because it is increasing all the time. Dense computer chips are used to make still denser chips and so on, doubling the number of microprocessors on a chip every 18 months, thereby doubling computer power. By 2015 it is expected that we will see a 137-billionfold increase of computational power compared with that generated by the first mass-produced model in 1959. As a result, everything is struggling to keep up: the human dimension in politics, society and economics, and also, of course, the human dimension in war. That's why we can say we think the future is making us rather than the other way around. 'When I pronounce the word "future"', writes the Polish poet Wislawa Szymborska, 'the first syllable already belongs to the past.' Or to quote the catchphrase of the US Air Force: 'If it works, it's already obsolete'.[19]

If we take both factors into account we can see just how far computerization is beginning to change our subjectivity. We ourselves are changing in relation to the machines we use. We are being trained to be more machine friendly. In business, for example, computers are now employed for 'informating', a process by which employees are given data in order to produce a better understanding of their market, to improve quality control and maximize market efficiency. In the military a similar process is at work, indeed it has been for some time. We can find the first evidence of this in World War II, when anti-aircraft gunners, submarine crewmen and radar operators were inserted into a mechanical and electromechanical system. They were seen, as many World War II poets saw themselves, as information transmitters and processing devices and their role gave rise in time to what is now called cognitive engineering. For not only does the military engineer hardware and software, it has begun to re-engineer its pilots.

In the 1990s McDonald Douglas issued a post-Gulf War advertisement for one of its new missiles, the Strategic Land Attack Missile (SLAM), referring to it as a 'man in the loop' guidance system. Today, the operator is increasingly absorbed into the weapons system he operates. The US Air Force (USAF), in particular, has amplified the cognitive processes of pilots to make them more efficient. In the case of the high-performance computer-based jet aircraft of today, pilots have to be capable of split-second responses. Given the continuous flow of information generated by onboard computers, their minds have to be made more machine friendly than ever before.[20]

The goal of military training is now increasingly design-oriented: to design operators who can process information even faster and thus interact with a computer environment more efficiently than they do at present. The aim is to improve, or enhance, human cognitive performance. The system is still in its infancy, as can be seen from an incident in 1988 involving the cruiser *USS Vincennes*. The *Vincennes* operated the Aegis Battlefield Management System, which drew on the latest digital computers and radar signal processing technology. Its manufacturers claimed it offered a comprehensive and infallible target system that could track more than 200 aircraft and missiles at the same time. Unfortunately, in 1988 the ship's commander was instructed by the system to shoot down an Iranian plane which turned out to be a passenger aircraft carrying 290 civilians. In other words, the system failed to distinguish between a 177 ft Airbus and a 62 ft F-14 Tomcat fighter. For its time, the Aegis system was more complex and advanced than any other, but it was in essence a man–machine system and most of the mistakes committed in the incident were manmade. Cognitive engineering clearly has some way to go before 'human error' can be eliminated.

Or can it? How possible is it to downgrade the human operator? Will we really see a day when the computers can think for themselves? What we do know is that computers have come a long way from Babbage's calculation machine. The speed and ramifications of calculation that war now demands may well shade into something different. The activities of computers may one day be closer to thought than calculation. The question then is, when will a computer designed and programmed by other computers outstrip not only human mental capabilities, but cognitive insights? Will computer systems in the future no longer require human operators?

Artificial intelligence (AI) enthusiasts have been predicting that this day will come sooner rather than later. All see 'expert systems', 'neural nets' and 'genetic algorithms' as incremental steps towards the ultimate goal of building machines that can think like ourselves.

- Expert systems are where specific areas of expertise of people's jobs are isolated and simulated in software.
- Instead of a single central processor, neural network machines involve a large number of tiny processing elements interconnected in parallel in a manner resembling the synapses that interconnect neurons within the brain. The hardware which runs these programs includes many processes running in parallel, allowing simultaneous processing of several different 'lines of reasoning' before conveying a solution.
- Genetic algorithms are intended not to write software in a final form but to allow it to 'evolve' through trial and error. They involve haphazardly throwing together snippets of a code in numerous combinations and putting them into an environment where only the 'fittest' (those able to reproduce) can survive. Random mutations are occasionally thrown in to ensure potential for further progress, in a way that is not dissimilar from the model of human evolution.[21]

Impressive though the attempts have been to allow computers to develop in an evolutionary manner, there is more to thinking than computing. Of course, we have designed computers that can extend our own abilities. But their processing power and hence understanding are strictly limited.

Indeed, although computers have been around for some time, those we use today merely eat up more memory and computing power; they don't do anything differently. The rate at which technology moves forwards depends not on the circuits we invent but on the software architectures we design. Software, not hardware, determines the pace of change. We can build ever faster circuits, new data transmission, and even more impressive hardware, but the software is still what it was in 1985. Computers cannot actually do anything differently from what they did then. They still have no flexible pattern recognition or language understanding, to name two of the broader human abilities that are central to our decision making. Until software is radically improved, computers will not be able to reason by analogy or produce testable hypotheses, both of which are central to second guessing an adversary's intentions.

Computers can indeed see for us, but they cannot work out the significance of what they see. A computer can scan the battle space and even recognize what its program asks it to look for, but it cannot understand what an enemy's deployment of forces might mean, or second guess its intentions, or intuit its battle plan.

This is the real difference between logical and intuitive thinking, between the use of algorithms (rules, procedures and instructions) and non-algorithms. Clausewitz called the latter 'genius'. He himself is taken as superior to von Bulow and Antoine Jomini, the algorithmic thinkers of their day. For he recognized that war is an art, not a science, and that it requires human creativity which computers do not possess. It is our creativity which defines our humanity.

If we don't always appreciate this fact, this is because, as I explained with reference to Gibson's *Neuromancer*, we tend increasingly to think of ourselves as machines, or we endow machines with humanity (comparing neural networks to brain synapses, and genetic algorithms to Darwin's struggle of the fittest). We think of ourselves as computers when we talk of 'programming' our lives and 'interfacing' with others, or 'processing' the information we are given, or insisting at times that it 'doesn't compute'. There is an inevitable tendency in the human mind, writes Hans Jonas, to 'interpret human functions in terms of the artefacts that take their place and artefacts in terms of the replaced human functions'.[22] For some time we have had a computational theory of the mind: the idea that our beliefs and memories are a collection of information, like facts on a database, and that thinking and planning are systematic transformations of these patterns, not dissimilar, in fact, from the operation of a computer program.[23] Wanting something or trying to achieve something else are merely feedback loops that execute actions to reduce the deficit – the difference, that is, between the goals we set and the actual state of the world.

But, in fact, reality is very different. For while we compare ourselves with computers, we think in a way that is different again from a computer. Contrary to the way we did in the past, when we talk of rationality or using reason, we do so not as the ability to think dispassionately, to reason out or calculate the 'maths', but in terms of our emotions – to reason our existence, to rationalize our irrational natures, to empathize and imagine the plight of others. We may not be the only animals on the planet that can reason or even compute, but we are the only ones that have imagination.[24]

So although like the mathematician Gödel we may talk of 'computable' and 'non-computable' thinking, we are not computers, any more than computers are like us. Indeed, what Gödel's 'incompleteness theorem' tells us is that even if we try to think logically, like a computer, not all problems are computable, or capable of being logically solved. As Gödel discovered, if you start with a set of logically consistent premises which lead to certain conclusions or consequences, then the power of logic will be unable to take you from those premises to all of the consequences and conclusions. We have to fall back on *imagining* what they might be.[25]

We can best appreciate this problem by looking at chess. Of course, a computer can now beat any human competitor. But war, like life, is not a game of chess. It is not predictable. Chess, in the end, is a game of finite combinations. It is actually a digital game and it should not surprise us that a digital machine can beat humans when playing it – why should we be surprised when they can eventually work through all the combinations, or through enough of them, at least, for the rest not to matter. War, however, is not logical. It is 'non-linear', to use the military jargon. It is unpredictable, the result of human passion. And it is unquantifiable as well.[26]

This is the main difference between digital and analogue thinking. Digital thinking defines reality precisely in terms of digits. Analogue thinking is approximate, or analogist. They are both ways of thinking but they are very different. Human beings deal in variables, computers in combinations. We reach decisions (or what the US Chiefs of Staff call 'decision dominance') on the basis of variables. We call this the processing of information and knowledge. But some of that 'knowledge' is already in our mind from childhood, from training, from what we have read or experienced, and all of it is independent of any particular 'game', including 'wargames'. No machine possesses the prior beliefs that are the essential ingredients of human knowledge, which is why at the moment no machine or computer can be creative.

There is another factor to bear in mind. Chess has no consequences. A player either wins or loses and moves onto the next game, or the next championship. War has very severe consequences which influence how we fight it. It is precisely because war has very real consequences for the human body that well into the current century we will still continue to see it as a *human* activity. Indeed, the future face of war will be shaped by biology, though our biological selves are in the process of changing or being re-engineered. This, as we shall see below, may indeed transform the soldier's subjective experience of battle.

Cyborg Warriors?

> The body is the first and most natural tool of man.
>
> Marcel Maus[27]

Even engineering is being transformed into a biologically based discipline. In the Massachusetts Institute of Technology Artificial Intelligence Lab, robots are assembled from silicon, steel and living cells. The activators of these simple devices are muscle cells cultivated in the laboratory, the precursors of the prostheses that will one day be installed seamlessly into disabled human bodies. Surgical body modification and biochemical alterations (for example through the use of botulinum toxin) are already commonplace. Within 50 years, or even earlier, these developments could be applied to enhance the abilities of tomorrow's cyborg warriors.

Manfred Clynes invented the term cyborg by mixing two words, 'cybernetic' and 'organism', to describe a subjective state which is both biological and technological at the same time. Clynes was a designer of physiological instrumentation and electronic data-processing systems, and he was the first to think of the cyborg as a 'self-regulating man-machine system'. The world's first cyborg was a white lab rat, part of an experimental programme at New York's Rockland State Hospital in the late 1950s. The rat had implanted in its body a tiny osmotic pump which, by injecting precisely controlled doses of chemicals, was able to alter some of the animal's physiological abilities. As a result the rat was part animal, part machine.

The Rockland rat was one of the stars of a paper which Clynes wrote in 1960 in which he contended that humans too could one day be modified to restore impaired functions (through prostheses and implants), or to enhance existing capabilities. Today a range of terms is employed to describe our evolving cyborg status, from biotelemetry to human–machine interfaces and bionics (the copying of natural systems).

'That there will be future warriors is the only certainty', claims Col. Frederick Timmerman, director of the Center for Army Leadership and former editor-in-chief of *Military Review*. His is a future which belongs to those countries, principally of course the United States, which by transforming the way technology is applied will be able to 'transform and extend the soldier's physiological capability'.[28] If war is to remain central to human culture, then soldiers' bodies as well as personalities may have to be reconfigured. In this regard, the cyborg condition has enormous implications for our humanity as well as our idea of war. For if endurance can be artificially enhanced, will we have to rewrite the ethos of the warrior, as man and machine continue to co-evolve equally?

Cyborgs, however, are not what sci-fi writers would like us to imagine. The popular view of the military cyborg can be found in Hollywood films such as *Robocop* and *Universal Soldier*. In the first the cyborg is a product of the Detroit Police Department, a subsidiary of OmniConsumer Products (OSP). The 'Unisols' (another industrial brand name) who appear in Robert Emmerich's 1992 movie *Universal Soldier* are cyborgs of a different stamp. Their hyperaccelerating bodies turn dead flesh into living tissue. The two principal combatants in the film, following their death in Vietnam, are flown back home to be packaged in ice, surgically eviscerated and refilled with cybernetic equipment, and thereby transformed into true twenty-first-century soldiers. A serum injected into the back of their skulls voids their memories. They are the ultimate killing machine with no sense of fear in the face of death, largely because, to all intents and purposes, they are already dead to themselves.[29]

What both films offer the viewer is a vision of a future in which technology is threatening to subvert the ontology of war as we have traditionally understood it. One obvious way in which technology has already modified our idea of soldiering is the admission of women into the armed forces, which represents a major break with 6,000 years or more of human history. In lightening the weight of weapons and providing integrated weapons systems that require less physical strength by their human operator, technology has removed one of the traditional barriers (physical as much as cultural) to the participation of women in war-related work. It will not be long before women soldiers will be permitted to take part in every form of combat, uninhibited by gender differences, whether biological or cultural.

It is precisely the fact that we will not be able to separate human and machine that explains the post-human condition. For the cyborg state goes beyond a partnership between man and machine; it is a symbiosis managed by cybernetics, the language common to organic and mechanical life. A classic account can be found in David Channell's *The Vital Machine*, which sees the future as a synthesis of two central currents of culture: the mechanized-organic worldview. For now organic systems are increasingly described in information-processing terms just like mechanical or informational devices. The line between the organic and mechanic is becoming blurred.[30]

And then there is Bruce Mazlish's *Fourth Discontinuity*, which talks of the co-evolution of humans and machines. Western intellectual history, he claims, has been marked by the overcoming of a number of great illusions (or what he terms 'discontinuities') which were once posited as natural and unchallengeable. The first was the distinction between humans and the cosmos (an idea challenged by Copernicus). The second was that between humans and other life forms (an idea challenged by Darwin). The third was that between humans and our unconscious (an idea challenged by Freud). And the last and most important distinction of all between the human and the machine is being overcome by technology.[31]

Now, it is important to recognize that this technology can take three forms.

- Restorative: restoring lost functions, replacing lost limbs and organs.
- Reconfiguring: creating post-human possibilities by adapting humans to their environment (Clynes's original work was on how to adapt humans to outer space).
- Enhancing human abilities, the aim of most military research.

Certainly, when it comes to restorative functions much progress has been made. But it is the third option which is probably central to the future of war, especially if we can escape Darwinian evolution.

For biocybernetics is not really intended to subordinate us to machines. Instead, it will help us use machines more effectively to enhance human performance. One popular sci-fi example is the way in which people interact with computers through computer chips inserted in their brains – the incorporation of silicon into their bodies. In William Gibson's cyberpunk stories data are transferred via 'wet-wired-brain implants'. These and other visions promise a world in which there will be a sophisticated interface between our nervous system and silicon; a world in which neural implants will enhance visual and auditory perception and interpretation, memory and reasoning; a world in which the distinction between computers and humans will be gradually elided.

Fibre-optic projectors can already throw images onto our retinas, thus allowing us to see directly without the intervening medium of a television or computer screen, and research is well underway to help enhance auditory senses through implants in the ear. The

USAF is investigating growing neurons in silicon chips to improve the communication between human and machine by allowing chips to be activated by hormones and neural electrical simulation. The Defense Advanced Research Projects Agency (DARPA) has a Brain–Machine Interface Program that, in its own words, aims 'to create new technologies for augmenting human performance through the ability to access non-invasive codes in the brain in real time and integrate them into a peripheral device or systems operation'. In plain English, this means enhancing human performance by working out how the brain controls movement and using the information to control external devices – transmitting (as has been done successfully with monkeys) brain signals over the Internet to operate a robotic arm hundreds of miles away. DARPA, which is affiliated to the military, shows the way to an age in which a warrior's brain may be part carbon, part silicon, and capable of using tools, including weapons, by the power of thought.[32]

For the moment, however, the future probably lies not so much in meshing machine and body as in fusing body and machine *functionally* rather than physically. Rather than integrate machines into our bodies, we will integrate ourselves into a system that will be shared between us. Functionally, we are already wired into digital networks such as the Internet which enhance our ability to process information. Some of the research projects already underway may be harbingers of the future. One of the most famous is the McDonald aircraft company's 'Pilot Associate', a project which has been ongoing since 1986. It is intended to allow 'expert systems' to evaluate the input from external sensors, as well as monitor and diagnose all the aircraft's onboard subsystems, including the pilot. Designed to predict what he will do next, it will be able to initiate actions if the pilot is unable to take decisions himself.

In our impending post-human future, we can glimpse a world that will force us to examine the question of what it means to be human. With the rise of cyborg technologies the existential dimension of war is being challenged as never before – above all, in the relationship between the warrior and his enemies.

'The primary political and philosophical issue of the next century will be the definition of who we are.' So writes Ray Kurzweil in *The Age of Spiritual Machines*. In so far as the new technologies promise to remake not only our bodies but also our worlds, they raise important and urgent questions about society's continued engagement with the soldiers who fight in its name.[33] And in so far as war is likely to be a struggle between the West and non-western peoples, states, societies, regimes or terrorists, its *intersubjective* meaning has never been more important, which is why even post-human warfare is likely to be just as ontologically real as before.

> Humanity is neither an essence nor an end but a continuously precarious process of becoming human, a process that entails the inescapable fact that our humanity is on loan from others, to precisely the extent that we acknowledge it in *them*. . . . Others will tell if we're humans and what that means.[34]

It is the idea of humanity as a process that brings us to the core of the question of post-human warfare.

In the future we will be encouraged to see humanity as a continuing process of 'becoming' human, a process that through cyborg enhancement (a form of 'participatory evolution') is now far more technologically determined than in the past. At the same time, morality is far more intersubjective than subjective. Which is why, of course, the prospect that we will fight post-human wars should prompt a sobering thought. In the man/machine symbiosis we might well glimpse an end of the warrior's alienation from his own humanity, but will we find him alienated from other human beings, those from non-western societies who are not experiencing the post-human condition?

Bio-enhanced Warriors

It is often said that science fiction is a genre of cognitive estrangement, a combination of the cognitive (the rational, scientific) and estrangement (translated as alienation from the familiar and the everyday). But, in fact, most science fiction writing is an extension or extrapolation of the present. If it were only concerned with estrangement we wouldn't understand it. And if it were only about cognition it would be a work of science, not science fiction. It is the combination of both that allows science fiction to challenge the ordinary, or what we take for granted.[35]

In Orson Scott Card's *Ender's Game* the training of soldiers in their early years takes the form of 'games' in the Game Room. The government has taken to breeding military geniuses and then instructing them in the art of war. Another influential work of science fiction is Leo Frankowski's *A Boy and His Tank*, which tells of a group of colonists on a planet combining virtual reality with tank warfare, a world in which warriors bond with their tanks, and their tanks with them. One of the most telling lines in the book is: 'Kid, if your tank is loyal you don't have to be!' Both books are read by the current American military. Card's novel is taught on the leadership course of the Marine Corps University at Quantico, and *A Boy and His Tank* was proofread by a solider from Company C Task Force, 1-32 Armour First Cavalry Division whilst he was deployed in the desert waiting for Operation Desert Storm.[36]

The Human Genome Project represents one of the most significant steps in our continued evolution. On one level it allows us to subject our humanity, which we have taken as a given, to biotechnological intervention. In theory, we could breed a warrior DNA or manufacture a race of warriors or 'natural born' killers. This is the premise of novels like *Ender's Game*. The object would be twofold: first, to make soldiers impervious to fear, fright or anxiety, and thus to make them more courageous (or foolhardy) in battle; the second is to make them more effective at killing. In pursuing the first object, of course, the second can be accomplished.

For the 'born' warrior is a killer, as well as one who is prepared – if necessary – to lay his life on the line. In her much-acclaimed book *An Intimate History of Killing*, Joanna Bourke writes that 'the characteristic act of men at war is not dying, it is killing', and it is as well to keep that always in the forefront of one's mind.[37] The soldier-killers analysed by Glenn Gray in his seminal study *The Warriors* (1959) are among the most formidable and terrifying warriors of all – men devoid of remorse or reflection, who exist in all armies at all times. Achilles was the supreme killing machine; so was the Alexander depicted so vividly in Arrian's history of Alexander's campaign. Indeed, what is so depressing in Arrian's account is that the slaughter goes on relentlessly. Clearly, its scale is what the author believes made his hero such a great figure. Killing is what Alexander did consummately well.

Is this a result of culture or nature? Evolutionary psychology argues that humans are born with a common set of preferences, predispositions and abilities fashioned by natural selection. These abilities enabled us to develop and become the dominant species on the planet. In other words, when we are born we are pre-equipped with abilities such as the ability to learn a language. We have something akin to a piece of computer programming that is designed specifically to enable us to acquire language skills. Experience – in the form of hearing our parents talk, for example – is just the input for the program.

It would also appear, however, that while we are all programmed to be aggressive, some individuals are genetically more prone to aggression than others. One of the reasons that war is likely to remain a male activity even in the mid-twenty-first century is that, across cultures, men kill other men 20 or 40 times more often than women kill other women. And the great majority of killers are of an age when the soldier is in his prime, usually between 15 and 30. It is also true, however, that in this same age group some are more inclined to kill than others. In western society, for example, 7 per cent of young men commit 79 per cent of repeated violent offences. If this is true of society in general, it must surely be true of the military, which, in a democracy at least, tends to be a microcosm of society as a whole.[38]

Of course, violence and war are not the same. A good soldier, Steven Pinker reminds us, is not the personality type that makes an intractable violent young offender: impulsive, hyperactive, with a low intelligence and usually an attention deficit as well. Young offenders are also resistant to discipline; they dislike being controlled. The worst of them are often psychopaths who lack a conscience and are more likely, if they join up at all, to be found in paramilitary organizations that set their own rules. Very rarely are they to be found in military units whose members are bound by close fraternal ties and tend, as a result, to enjoy a high degree of self-esteem.[39]

One of the interesting phenomena of war is the extent to which a very small percentage even of professional soldiers actually kills with any real enthusiasm. It has been calculated, for example, that 1 per cent of fighter pilots accounted for at least 35 per cent of enemy aircraft kills in World War II. Clearly, they were not only more talented but more aggressive than their counterparts. On the ground the figures are even more remarkable. Take, for example, Lance Sgt Simo Hyha of the Finnish Army who in three months in the winter war of 1939 killed 219 Soviet soldiers with a standard-issue service rifle. Another born killer was Sgt Alvin York of the American Expeditionary Force in World War I (memorably portrayed in film by Gary Cooper) who killed 28 Germans in the battle of the Argonne on one day alone, 8 October 1918, a month before the Armistice. Looked at differently, York single-handedly accounted for the equivalent of two German companies. For as well as killing 28 soldiers he captured another 182; his exploit could potentially have transformed a tactical situation along a key sector of the front line.[40]

Killing, in short, doesn't seem to come naturally in all situations to all soldiers, even the most highly trained. Natural soldiers are not made, they are born. And there are very few of them. That's why the military has preferred the discipline of collective units such as gun crews, which are more easily controlled and which, being often distant from the battlefield, are also less emotionally involved – in a word, more 'mechanical'. The greatest cruelties have been the impersonal ones of remote decision, system and routine, especially when they can be justified as operational necessities.

So, looking at the situation from the perspective of the evolutionary psychologist, we can sum up as follows. We are not born to kill any more than we are born to engage in war. Killing is a contingent strategy connected to complicated circuitry that allows us to compute subconsciously whether it is in our interest to kill or not. We deploy aggression as a strategy and we do much less so than we have done before. Like war, violence has declined among the rich nations, not because of morality or ethical codes. Ethics tends to legitimize, after the fact, the contingent strategies we have already chosen; it is cultural and learned. If we have become more ethical (in our own eyes, at least), it is because we have become more cosmopolitan (or less hostile to strangers). The technologies that promote literacy, travel and knowledge of history have all contributed in different ways to that growing cosmopolitanism. Our social imagination has expanded as a result. Through television and film we are able to project ourselves into the daily lives of other people, even though they are remote from us both physically and sometimes emotionally.

Of course, if our postmodern societies continue to discourage violence or sublimate it through sport, then they may find themselves with a smaller pool of talent from which to draw natural born soldiers. Thus we may have to look to pharmacology to stay in the war business. Drug enhancement rather than genetic engineering is likely to be the key. For it is far too early to talk of the genetic engineering of soldiers, something which, if it happens at all, we are more likely to encounter beyond the half-century.

One way of doing this, popularized by science fiction writers, is cloning – the transplanting of a mature human cell with its full DNA pattern into another human egg whose nucleus has been removed. Cloning is the way to transmit the genetic signature of one parent to an embryo and thus effectively to create a genetic identical twin of the parent. This has given rise to the fear, elegantly summarized by Richard Dawkins, of 'phalanxes of identical little Hitlers goose-stepping to the same genetic drum'.[41] But current opinion is that the cloning of human beings will remain genetically difficult, if not impossible, for years to come. Instead, cloning is more likely to be used to provide cell banks for the living, that is, replacing parts lost on the battlefield. The future of cloning probably lies in spare part surgery rather than the replication of human beings.

A much more profitable subject of speculation, because it has been happening for some time, involves the modification or control of human behaviour through neural pharmacology. In the 1980s the US Army study of future war, *Battle 2000*, stated that 'battle intensity requires stress reduction'. Every army has attempted to reduce stress and thus improve combat performance by whatever means come to hand, alcohol being the oldest. Today, through continuous advances in neuroscience, it is proving cheaper, easier and much more productive to control anxiety and fear through pharmacological means. In civilian life there is a host of drugs now available to improve performance – sexual (Viagra), sporting (steroids) – and even to boost self-esteem (Prozac). Where the civilian leads the military is not far behind.

In late 1944 Nazi doctors tested a cocaine-based drug that they hoped would improve the performance of their war-weary troops. The pills were given to inmates at the Sachsenhausen concentration camp who were forced to march 55 miles carrying 20 kg packs. Germany lost the war before the pill could go into mass production.[42] Today scientists are much more sophisticated. Even in democracies amphetamines were dispensed to British Forces in World War II. The Vietnam War, the first 'pharmacological war' in history, saw 50 per cent of soldiers take drugs, either those prescribed by the authorities in the form of anti-depressants or those regularly traded in cities such as Hue or Saigon where soldiers openly self-prescribed marijuana and heroin.[43]

As for airmen rather than ground combat units, the American military continued to give amphetamines to pilots on transcontinental missions in the 1950s and 1960s as well as in the first Gulf War. Amphetamines and tranquillizers – 'go' pills and 'mogo' pills as they are popularly known – are considered useful tools for a modern American military that likes to fight at night, given its technological superiority in finding targets in the dark, and for an Air Force that must expect its pilots to fly longer missions from fewer overseas bases. Scientists are researching even more potent pills, including some that keep combat forces alert for 40 hours or longer.

In the immediate future the military will probably also try to manipulate the endogenous opiate system in an attempt to decrease sensitivity to pain, and thus enhance physical stamina and mental endurance. The genetically modified soldier is part of the Pentagon's search for an 'Extended Performance Warfighter', a program that focuses on using devices other than drugs to enhance performance. TMS or electromagnetic energy may allow scientists to 'zap' a soldier's brain, giving him the capability to stay awake, fight and make decisions for a week. In the future wearable devices attached to clothing may also be employed to gauge a soldier's mood by the number of eye blinks. Internal implants able to monitor the heartbeat may be capable of administering tranquillizers or sedatives without the soldier knowing.

There may well be a price to pay for all of this. Francis Fukuyama talks about a trade-off that faces all of us at the beginning of the twenty-first century: the constant pressure to reduce the ends of biomedicine to utilitarian ones and the lesser pressure to hang on to our humanity – or, in the case of the warrior, his existential status. Is the post-human condition one in which our humanity will be defined in terms of single categories such as pain and its relief at the cost of what pain can often produce: creativity in the arts, and endurance in war?[44]

There are clearly important subjective implications if the military are to continue to go down this route. For the prevalence of some 'disorders' such as fear and anxiety can be traced back to the fitness benefits that the predisposing genes confer. Fear and even guilt can be seen as adaptations and defences which serve a useful function. Flashbacks and the recurrent memories of post-traumatic stress disorder (PTSD) sufferers arise because the mind finds it useful to remember life-threatening dangers in order to avoid encountering similar dangers in the future.

One day we may indeed be able to eliminate pain, but will we do so at the cost of ambition and genius, the ends that are often set by people in pain as a therapeutic strategy? Nietzsche argued that if we are in pain we should not try to eliminate it so much as to put it to creative use. No one these days would be in favour of Nietzsche's views (which were conditioned in part by his own suffering), especially now that we have a chance to reduce pain, or even eradicate it completely. But some of the most admirable human qualities are often related to the way we deal with pain and suffering, and ultimately death (something else we are trying to postpone for as long as possible). In the absence of pain, will warriors find in themselves the same degree of sympathy, compassion or strength of character they were able to find in the past?

A different question is raised by mood-altering drugs that may make us more self-confident or brave. For there is an immense difference between self-esteem that is earned or 'won' – the term itself is highly revealing – and the self-esteem that is produced independently of action or, in this case, battle. This is a matter that I discussed in the last chapter.

A third question is more pointed: is war about to become a solipsistic exercise for the first time? Hobbes described the state of nature, in a phrase that has echoed down the centuries, as 'nasty, brutish and short'. Often, we forget the fourth word he used: 'solitary'. Whatever war was in the past, it was rarely a solitary activity. Indeed, the popularity war has had for soldiers has come largely from comradeship, from the fraternal links it promotes. Will this continue to be true of post-human warfare? Such camaraderie is often dismissed as male bonding, but it is real enough. The problem with some drugs is that they produce a high (or a condition) which is incompatible with any kind of typical social functioning. Drugs often promote introspective or self-referential behaviour, and in that sense may be detrimental to the primary group cohesion that the military values so much.

Conclusion

All of which raises an important question about the existential dimension of war. Biotechnology may enable soldiers to become mentally and physically tougher by means other than the traditional ways of example, teaching or tradition. Let us take courage as an example. Traditionally, courage was the virtue of the warrior class which corresponded, for Plato, to the spirit in the individual. It was also an existential virtue, a celebration of one's own existence. What made it distinctive was that it was born of the moment. Some were born brave, others discovered their bravery in the heat of battle. To show courage was to be true to one's own being, which is why it was existential, as Socrates argues in one of the Platonic dialogues, *The Laches*. The man who enters battle fearfully and is uncertain of its outcome has more courage than one who is confident of victory from the start. If there's no fear, in short, there's no courage. For courage is the overcoming of fear. It is the ultimate 'paradigm for self-confrontation'.[45] But it is also a culturally constructed one. In terms of 'excellence', the soldier shows courage to win glory or distinction in the eyes of others. In terms of 'merit', he shows courage not because it is in him but because he expects it of himself.

But what would happen if, in future, courage is biologically determined or at least biologically enhanced? Librium and Valium already treat anxiety, Prozac and Zoloft depression. Prescribed drugs are regularly given to Air Force pilots to reduce stress and fatigue and enhance wakefulness for up to 72 hours at a time. Anxiety suppressants are now given to pilots going into combat. Viagra is given to the SAS to boost testosterone levels and thus aggression. The increased geneticization of soldiers and pilots threatens to undermine the traditional culture of service, discipline under fire, as well as endurance, which is usually the product of training, or *esprit de corps*.

And what would happen if we could abolish guilt and thus neutralize the sometimes traumatic consequences of showing courage? What if, by swallowing a pill, a soldier could immunize himself from a lifetime of crushing remorse? For the prospect of a soul absolved by medication is about to become real. Feelings of guilt and regret travel neural pathways in a manner that mimics the tracings of ingrained fear; thus a way of addressing one should address the other. Experiments have been conducted at the University of California at Irvine to inhibit the brain's hormonal reactions to fear, softening the formation of memories and the emotions they evoke. The beta-blocker Propranolol nips the effects of trauma in the bud. It is proving possible to short-circuit the very wiring of primal fears.[46]

Another research team at Columbia University has discovered a gene behind a fear-inhibiting protein uncovering the traditional 'fight or flight' imperative at a molecular level. Will we soon be able to blunt the human conscience; to mediate out of the psyche regret, remorse, pain or guilt? Is this the ultimate end of 'consequence management': making the soldier blind to the consequences of his or her own acts? Is this the thin end of a dangerous wedge: the emergence of a morally anaesthetized soldier?

4

Beyond the Death Threshold

War and the Spiritualization of Cruelty

If the new biotechnologies make it possible to enhance human abilities, they may also add to the ethical dilemmas of war, or even transform them. Whatever the new age may produce, it will magnify the capacity of human beings to act as they always have done in the past. If we are to continue to fight wars, we must ask whether we can do so in an enhanced fashion by being more cruel (or cruel in different ways), or less cruel. Is cruelty essential to war, or is it a historically contingent factor, a product of a technology that was far more destructive in the past than it is now, and a warrior elite that once exulted in cruelty as an expression of its own 'will to power'?

Mention of this phrase inevitably brings to mind Nietzsche, and we might perhaps begin this chapter on ethics by asking, as did Nietzsche himself, whether we could produce a historical anthropology of cruelty. In an aphorism entitled 'Something for the Industrious' which appears in his book *The Gay Science*, he recommended all kinds of new historical investigations for the histories 'of love, of avarice, of envy, of conscience, of pious respect for tradition or of cruelty'.[1] For cruelty he believed was endemic to life. One could argue that philosophy began with the question, why pain? It is at the basis of every religion, every account of a fall from a golden age and the invention of hell, the kingdom of pain-as-punishment. It is at the heart of the discovery of sacrifice. Because we are embedded (i.e. we have a body that decays and dies), we needed to invest both pain and death with social meaning. It is in the act of investing our lives with meaning that we become what we are (that we escape from a 'being which is pain' into a being who suffers for a reason).

Now, although inflicting pain on others has always been a central feature of war, frequently cruelty has been disguised. 'Almost everything we call "higher culture" is based on the spiritualization of cruelty, on its becoming more profound', wrote Nietzsche in *Beyond Good and Evil.*[2] And he attributed this largely to aesthetics. In the Bronze Age the warrior elite had been 'protected by the hand of Homer', who with 'artistic deception' had consecrated the 'unspeakable ugliness of war'.[3] Every other age had taken its cue from ancient Greece's greatest poet.

Nietzsche did not mean by this that the cruelty of war had quite literally been disguised, although it was often true that the reality of war – the injury one side deliberately inflicts on another – has been omitted altogether from many conventional accounts. We are still anaesthetized to the dead through the use of euphemisms such as 'the fallen'. Euphemism and omission are the two main ways in which war has been sanitized through the centuries. Rarely in the past 300 years do we find graphic accounts of death, and even fewer of mutilation.

But if we care to look for it, cruelty can be found on almost every page of the *Iliad*. Homer's depictions of battle are as gruesome as any post-World War I narrative: the spear that cuts through the sinews of Pedaeus's head, passing through his teeth and severing his tongue, or the bronze point that enters Phereclus through his right buttock, piercing his bladder and bone. 'The funeral pyres burned night and day', Homer reminds us, lest we think that the Trojan War was carried out entirely in episodic encounters between heroes like Hector and Achilles, Ajax and Paris. It involved death by the thousands. What we find in the *Iliad* is not a battlefield so much as a killing ground on which carnage is the prevailing reality.

Centuries later the Greek tragedians exposed the brutality of the Trojan War even more graphically. In *Ajax* Sophocles presents us with the first picture we have in the ancient world of a traumatized victim who, unable to live with himself, eventually takes his own life. In *Philoctetes* he portrays a disabled soldier abandoned by his friends on their way home because they cannot stand either the stench of his festering wound or his screams of pain. 'Terrible it is beyond word's reach', Philoctetes says of his condition. The unspoken aspect of human pain is what dreams try to voice – pain so intense that it reduces life to emptiness. As Aristotle recognized, pain can be so all consuming that it destroys even the nature of the person who experiences it.

In fact, as Nietzsche tells us, what is remarkable about the Greeks is that they went out of their way to celebrate the cruelty of war. 'The delight in making suffer becomes an *enchantment* of the first order', he wrote.[4] Enchantment derived from the ability to impose their will on others and, in the case of endurance in adversity, to impose their will on themselves. Nietzsche began life as a philologist, or student of Greek literature, and he set out to challenge the received idea of his day that the Greeks had celebrated only beauty and harmony in their art. For he recognized that they had also celebrated its Dionysian elements – the frenzy and brutality of battle in which they exulted. What Nietzsche understood was that suffering – or making other people suffer – as a celebration of their own will to power is what societies in the next 20 centuries were to find particularly enchanting about war.

Indeed, what was most disenchanting about modern warfare was that society could no longer celebrate war in the same fashion. After 1870, we had to confront more directly than ever the problem of human pain because it could no longer be disguised.

Take physical mutilation. It may well have been a reality in the past but it was not a major, or even minor, theme of poetry or tragedy, let alone art. There were a few exceptions such as Jacques Callot's seventeenth-century depictions of *The Large Miseries of War* and Goya's even more graphic depiction two centuries later of *The Disasters of War*, the atrocities committed by both French soldiers and Spanish partisans during the Napoleonic Wars. As befits a baroque artist, Callot expressed emotions in formal terms even when depicting a woman being raped in the presence of her husband, or a man vomiting on a heap of corpses. Goya, by contrast, looks forward to the future in showing that while war brings out goodness and courage in some, it may also reveal the worst in everyone. The fact that he showed the partisans in a not altogether flattering light may explain why his pictures were never released in his lifetime.[5]

No such reticence distinguished the Great War poets, who took their inspiration from Charles Sorley's injunction:

> When you see millions of the mouthless dead
> Across your dreams in pale battalion go,
> Say not soft things as other men have said.[6]

Sorley died early on, in the battle of Loos (1915). His successors had to find a language for the things they had seen, and they did. Wilfred Owen wrote in an unfinished preface to his poems, 'All the poet can do is to warn', and one of the most telling warnings was to show what happened in war to the human body.

What was especially disenchanting about the industrialized battlefield of the twentieth century was the horror of mutilation. The public veneration of the 'Unknown Soldier', the poor wretch whose body had been the most disfigured and the most completely broken, the one no longer recognizably human, was its sole virtue, for being emblematic of what war had become.[7] When at Verdun a group of soldiers took a direct hit, the result was surreal in its horror: 'the great power of earth, round, shaped like a pyramid, with a hole gauged out all round. Sticking out of it, symmetrically, to a distance of about 40 cms, were legs, arms, hands and heads like the bloody cogs of some monstrous capstan.'[8]

The industrial metaphor was well chosen given that the new horrors of war were mostly, though not entirely, industrial in origin.

In the Great War high-velocity bullets and shrapnel both produced severe wounds. Despite steel helmets, 10 per cent of all injuries were to the head. In response plastic surgery developed into a leading aspect of medical science. This was the disenchantment of industrialized warfare – the scandalized vision of death: the disembodied body, the atomized victim. This applied equally to the aircrews of World War II, especially the tail gunners on the great Flying Fortresses like the B17 who were especially exposed to danger, even in the air. In the words of the American poet Randall Jarrell, most were as powerless as a rivet in a ball turret; most were 'reduced to a neuter'.[9]

By then cruelty could no longer be 'spiritualized', only shown. Even the war artists of the twentieth century usually turned a harsh but necessary light on experiences that the generals and politicians would have preferred to disguise, or to 'spiritualize' away. This is as true of Paul Nash's World War I masterpieces such as the corpse-strewn landscape of *We are Making a New World* as it is for Otto Dix, who showed soldiers suffering a thousand deaths in combat. Even those who survive (one thinks of the mutilated men in *Card Players*) fare little better, for they are trapped in bodies beyond repair. Leading a shadowy existence they are overwhelmed by the 'fleshiness' of the civilian figures, or find themselves juxtaposed to other civilian victims (most memorably, the veteran and prostitute in *Two Victims of Capitalism*, the veteran with a hole for a mouth, the prostitute whose syphilitic facial sores look like bullet wounds).

Mutilation also encompassed nature. For warfare had become environmentally disenchanting as well. Dix's *Trench with Barbed Wire* illustrates vividly the degree to which the earth has been 'disfigured'. Dix seems fascinated by what one critic, in potent metaphorical language, calls 'the surreal dismemberment of the earth', and another its 'mutilation'.[10] Paul Nash's battlescapes of blasted nature are equally stark in form: brutal, coarse, raw in design, with no sunrise or sunset (no beginning or end), a land disfigured by great gashes, a panorama largely devoid of men. But the apparent absence of men, of course, subliminally draws our attention to their presence – in the trenches, sheltering from the bombardment like human troglodytes. Ernst Junger wrote that he had spent four years 'in the midst of a generation predestined to death spent in caves, smoke-filled trenches and shell-illumined wastes'.[11] His tragedy and that of his comrades was to inhabit an industrialized battlefield in which men could no longer live in harmony either with nature or with themselves.

What is interesting, however, about the disenchantment of war in this period is the fact that, apart from artists such as Dix, others found it increasingly difficult to depict the cruelty of war because of developments in their own profession. One of the chief sources of the disenchantment of modern life is what T. S. Eliot called 'the dissociation of sensibility'. Art lost touch with life. By 1890 the artist was more important than the work of art. The work of art merely reflected the relationship of the artist to the world, including his own inner thoughts, sensibilities and perspectives.

Perhaps the last war picture to have immediate political impact was *Guernica*, Picasso's savage indictment of the Spanish Civil War. It depicts a horrific moment: the bombing of a Basque village on 26 April 1937 by German and Italian pilots in support of the local Fascist cause. Even today the painting has the power to evoke the horrors of war, provided, of course, you know what you are looking at. As the bombs plummet, one woman flings her arms up in a despairing attempt to stop them. A mother cradling a dead child on her lap lifts her head and parts her mouth to scream. Over an injured horse, a naked lightbulb sends out shafts of brightness. But the gutted animals and screaming mothers might have been expressing any of the many other horrors of the twentieth century, the concentration and death camps, as well as the man-made famines. The painting might not be about war at all. This is why photography rather than pictorial art was more decisive in capturing the reality of war, largely because it performed a social function that not even artists such as Picasso were able to do.

What makes photography intensely social is that it is representational, and as such offers us more than the photographer's understanding of reality. To be sure, a photograph is an art form because it offers us a subject that the photographer chooses to capture on film. Just like a modern pictorial work of art, a photograph highlights the relationship between the subject and the photographer. Like the artist, the photographer does not offer us the whole truth, merely an interpretation of it. But André Breton called the camera 'a savage eye' because it presents more than a photographer's understanding of what is real. The most powerful photographs capture the real, unadulterated by the rationalization of art. Since 1905 a painting has always told us more about the artist than the subject itself, largely because photography took over the main responsibility for representing reality. We always ask first who painted a picture; we rarely ask who took a photograph because a photograph can stand independently of the artist who took it.

At this point we must distinguish between the still photograph and the cinema. The progressive element of art was limited by its very medium. It remained obstinately the same, as did the materials of the artists. The greatest limit to the pictorial representation of reality (and this is true of still photography as well) was that it could not show *movement*.[12] Once this was recognized the medium itself had to be transformed, as it was at the end of the nineteenth century with the coming of motion picture technology. After 1905, when people wanted to see the reality of modern war they went to the cinema. In Britain audiences could watch reconstructions of the Napoleonic Wars (*The Battle of Waterloo*, 1913), the American Civil War (*The Battle*, 1911) or even the Anglo-Boer War as it was taking place (*The Call to Arms*, 1902). They have been watching ever since.

Now we arrive at an interesting question. Has the moving picture blunted the 'spiritualization of cruelty' or has it aided it, not by supplanting pictorial representation but by *transcending* it? For the main aesthetic of today's wars is speed, a process in which the individual soldier, not the artist, finds himself caught up. As Milan

Kundera puts it in *Slowness*, 'Speed is the form of ecstasy the technical revolution has bestowed on man'.[13] The aesthetic of war is being captured by the cinema, more recently by the video, and now by the computer-generated image. This raises the moral question, to what extent does the camera do the seeing for us? Are we beginning to see war from the machine's point of view?

The philosopher who first suspected this was already the case was Martin Heidegger. In the cinematization of war and its speed, or its 'motorization', Heidegger believed technology itself had become metaphysical for the first time. Referring to films showing the German occupation of Norway in 1940, he observed:

> When in order to capture the intrepid image of airborne troops jumping from planes an additional airplane is called in to film the paratroopers, there's nothing 'sensational' or 'curious' about it. Such effusion of these images is in and of itself part of the event. Such 'film reportage' is a metaphysical process.
>
> What we have here is technology as sensationalism.
>
> From the perspective of 'spirituality and bourgeois culture' one tends to consider the complete 'motorization' of the Wehrmacht . . . is boundless 'technicalism' and materialism. But it is really a metaphysical act.[14]

What Heidegger was discussing was metaphysics as 'the will to power' and technology as the manifestation of that will in the late modern age.

Is this even more true today? For the total air superiority which the US Air Force (USAF) has achieved since the late 1970s has made war into a routine. For the pilot war itself is largely cerebral, not visceral. If it is visceral at all, does its attraction lie in the will to power? In the first Gulf War US bomber pilots flew missions with heavy-metal music pumping through their headsets while graphic-simulated displays helped guide their bombs to their targets. In Afghanistan in 2002, as Mark Bowden reminds us, pilots were even able to get home in time to catch the latest episode of *Friends*. In his report from the front he recalls the following incident:

> Among the squadrons' recorded collection of audio-video 'greatest hits' was the artful destruction of the purported Taliban building in Kandahar. Last summer I reviewed the event with a group of crew members at their base in Idaho. On the monitor we even watched a negative black-and-white thermal image of a building at the centre of the city. Vehicles and people were moving on the street out front. Abruptly four black darts flashed into the picture from the upper-lefthandside, quickly as an eye blink, and the screen was filled with a black splash.
>
> On the recording the gleeful voice of a wizzo named Buzzer shouted 'Die like the dogs that you are' . . . On the screen in the form of tiny black dots people could be seen emerging from the flaming building, fleeing down the street.[15]

The question inevitably raised by incidents such as the above is what the camera is showing us. That is why aesthetics is central to ethics. For aesthetics does not just relate to art but, as its Greek root (*aesthesis*) suggests, to a whole region of human perceptions and sensations. As Nietzsche tells us, 'Life itself is "aesthetic" because it aims only at "semblance", "meaning", "error", "deception", "simulation", "delusion" and "self-delusion"'.[16] Representation is not reality, it is a version of it. It tells a story – but who is telling it? This is the question around which the ethical debate on war now largely revolves. And it is perhaps the key to whether war can successfully be re-enchanted.

For Hans Jonas, writing in the 1970s, it was especially important that there should be an 'ethic of responsibility' in an age when we are distanced from our actions, both geographically and over time. Jonas spent much of his life warning that modernity was redefining the parameters of human action. Traditional ethics saw action in terms of its immediate effects on others who were usually close at hand. In the atomic age we had to be much more aware of the long-term effects of human behaviour on those remote from us in space, and even in time (future generations, like the children of the citizens of Hiroshima, many born with genetic defects as a result of their parents' exposure to radiation).[17]

In other words, in the re-enchantment of war we find an ethical dilemma as well as an instrumental one. What adds to the ethical dilemma is not the computer itself (there is no reason why war in the information age should be any less ethical than the industrial-age model). The problem the computer brings should be seen as somewhat different: speed, and speed, in turn, as an aesthetic. Susan Jeffords writes of 'the aesthetics of technology' as 'a fragmented collection of disconnected parts that achieve the illusion of coherence only through their display as spectacle'.[18]

Not that this is entirely without precedent. As early as the 1930s Walter Benjamin wrote about the triumph of the modern age in pushing death from the perceptual to the *conceptual* world, in removing it from the home and from everyday life and 'sanitizing' it behind the walls of the hospital or sanatorium.[19] But something much more complex is happening in war. The new technologies have drawn the military into a world in which reality is mediated or simulated. This is one of the ethical dilemmas of the American way of warfare because it creates an encounter that is qualitatively different from before – one that has little or no reference to ourselves. What the warrior sees on the computer screen is also real. The perception of reality becomes reality. The medium, as Marshall McLuhan famously proclaimed, becomes the message. It assumes a life of its own.

Perhaps that is the future, too. If the true purpose of re-engineering the human condition is the abolition of obstacles to the reproduction of machines, then obviously we must become more machine friendly. We must be persuaded to see the world from the machine's point of view even if this means a simultaneous hardening of the human heart. Distance is in danger of being transformed into disassociation, given that the form it takes is technological (or instrumental) in two critical respects. The first is the mental disassociation between the cruise missile operator and his target: pilots or naval operators are increasingly cut off from the consequences of their acts. Secondly, reality is increasingly mediated by the computer, which does the seeing for us. As a result, today's pilots are increasingly cut off from responsibility for their actions.

But what is interesting about distance – especially the new distance of cyberspace – is that the first action on our part is to place ourselves firmly within it, not to stand apart. William Gibson, who coined the term cyberspace, clearly recognized it as a non-space but chose to represent it in spatial terms, striking a chord in the minds of the rest of us that still echoes today. Internet browsers still *feel* that they 'visit' a website and that they are 'surfing' the Net, while members of discussion groups or online communities *feel* that they are sharing a common arena (sometimes spatially represented through graphics).

The computer, in other words, and the information revolution of which it is the central part are not necessarily divorcing us from our own actions. The picture is more complex, and in some senses more encouraging. To understand why, we must go back to history and ask why the computer was first pioneered by the United States.

The United States and War in the Late Modern Age

Traditionally, the Americans have always preferred the instrumental dimension of war, so much so that they have gone further than anyone else in attempting to *instrumentalize* it. The first attempt came in the late nineteenth century and was a reflection of a modernity peculiarly their own. The second (the present attempt) is an expression of a postmodern condition. But what makes war postmodern is its increasingly instrumental nature. What is particularly ironic is not that the United States should have been the main country to have disenchanted war (inadvertently to be sure), but that it should be the only country now trying to re-enchant it instrumentally.

By 1860 the United States had outstripped every other country in the development of machines to do jobs previously undertaken by skilled workers. In *The Social History of the Machine Gun*, John Ellis advances several reasons to account for this engineering preeminence. We should remember Heidegger and his insistence that the essence of technology is not technological:

> The manufacture and utilization of equipment, tools and machines, the manufactured and used things themselves and the needs and ends that they serve or belong to are what technology is.

Ellis provides three reasons for America's fascination with acquiring machines to replace skilled labour. The first was an acute shortage of manpower, hence high wages. To keep prices down, productivity had to increase. Machines and rationalized, centralized production units were introduced to multiply the productivity of the individual worker. Secondly, the United States lacked a well-organized class of handworkers who would consider mechanization as a threat to their way of life. One of the pioneers of mass production, Eli Whitney, explained its purpose as 'to substitute incorrect, inefficient operations of machinery for the skill of the artist which is acquired only by long practice and experience'. Lacking such skills, American industrialists pioneered what they called 'a new way not of making things but of making machines that make things'. Thirdly, the need to think in terms of mechanical possibilities as opposed to the limits of human skill threw up a new set of experiences concerned solely with designing better machines. In the country's fascination with mechanization, machines, not men, became specialized.[20]

The United States also pioneered the world's first machine tool industry. Machinists of the Franklin Institute and their colleagues in cities like Cincinnati and Providence, Rhode Island, built lathes, planers and drills so that US industry could turn out rifles by the million. An important figure was William Sellers, who persuaded the US Navy to move from screws, nuts and bolts that were custom made by machinists to a national standard. Today there are 800,000 global standards; in the 1860s there were none.[21] The upshot of all this was that the United States can be said to have introduced a central instrumental/mechanistic principle into war: the serialization of death on the battlefield.

But US history also affords a second example of instrumental thinking: the instrumentalization of the labourer – including the soldier on the industrialized battlefields of the twentieth century. For the United States was the first society in which men were geared to machines, in which the labour force was synchronized for the machine's more efficient use. American workers were expected to work with machine-like precision. As we have seen, in the first years of the twentieth century Frederick Winslow Taylor wrote at length about how the body could be treated as a machine. As such he tended to treat it not as a human entity so much as an inanimate object to be exploited and directed at will, according to specific laws revealed by natural abstraction.[22]

Taylor was linked with the scientist Thomas Edison and the manufacturer Henry Ford. All three helped instil capitalism with a fierce obsession with time, order, productivity and efficiency. Lenin was drawn to Taylor's vision of a conformist, consensual, planned society. So too were American liberals. In the 1950s the management guru Peter Drucker listed Taylor with Freud as the twentieth century's most important thinkers.

Thirdly, the United States tried to instrumentalize war by forging a new partnership between society and science. Modern American scientists were systems builders. A particularly good example is Edison. He was a quintessential American scientist, for not only did he invent the electric lightbulb, he also pioneered the development of electricity generation from transmission to the home to distribution within the United States. In doing so, he had to be practical in estimating the cost of laying the cables, running power stations in the cities and looking for commercial franchises, one of which was the electric chair. He had to sell the idea of electricity as a way of replacing old methods in that most traditional of businesses, capital punishment. In a word, there was no division between the scientific and the social. What appeared to be social was now technical. What appeared to be technical was now social. The United States saw the birth of what historian Thomas Hughes calls a new socio-technical order. What emerged was a *technicist* ideology promising a technological fix to every military problem.[23]

In a postmodern, post-industrial age, the United States has reinvented war once again, in a way that is consistent with its cultural preferences. It is an instrumental choice that is consistent with past practice in all but one crucial respect: technology offers it a chance to reduce the cruelty of war. In displacing the mechanistic ethos of war, it has allowed the fate of the human body to become of prime concern.

Despiritualizing Cruelty

It is this focus that may allow war to retain its ethical appeal. This becomes particularly clear if we look at the computer, as perhaps we should, in terms of its implicit rejection of metaphysics and its embrace of its biological possibilities. For the computer, contend George and Meredith Friedman, has become the definitive mark of the American system. What makes it definitive is its pragmatic character. The computer doesn't contemplate aesthetic, moral or ethical issues. Instead, its programming language focuses on solving immediate and practical questions. To that extent, it expresses the American spirit, one that finds little time for metaphysics.[24]

Even Nietzsche might have been happy. For after announcing the 'death of God', he observes that his interpretative shadows linger on in the great metaphysical systems that came to dominate political life. And then he asks a very postmodern question:

> When will all these shadows of God cease to darken our minds? When will we complete our deification of nature? When may we begin to *naturalize* ourselves?[25]

Nietzsche thought that 'the naturalization' of man ought to be one of the central ideas on the agenda of philosophy. Indeed, he proclaims that one of the many tasks of the 'philosophy of the future' will be 'to translate man back into nature':

> To see to it that Man stands before Man . . . death to the siren songs of old metaphysical bird catchers who have been piping to him all too long. 'You are more, you are higher, you are of a different origin!'[26]

What Nietzsche is interested in is what makes us human – our biology. He won't even hear talk of the soul. As Zarathustra says, 'the "soul" is only a word for something about the body'.[27]

Unfortunately, instead of revaluing the body, the modern age revalued the spirit. Marxism and Fascism were a case in point. Fascism dealt in the language of the spirit, especially the 'will', Marxism in the language of matter, but both privileged culture over nature; both taught that human nature could be improved through political action. And both believed in being cruel to be kind. Orwell captured the schizophrenic nature of his age very well. Practices which had been long abandoned – the use of prisoners of war as slaves, torture to extract confessions, the taking of hostages and the deportation of whole peoples – were not only common once again but tolerated and even defended by people who considered themselves enlightened and progressive. This was an age when 'human equality becomes technically possible'. Cruelty in the name of 'human equality' is what made the twentieth century unique.[28]

In explaining the rejection of creeds such as Marxism, the poet Octavio Paz talked of the '*revolt of the body*'.[29] It is a telling phrase. For have we not *re-enchanted* life by rediscovering the sacred in ourselves (not in science), and done so in part by robbing science itself of its 'metaphysical' status? Have we not begun to rethink and reformulate our relationship to nature, and feel we have been forced to do so because modern science's mechanistic and objectivist conception of nature was not only limited but fundamentally flawed? This is suggested at least by the new approaches associated with Bateson's 'ecology of mind', Bohm's 'theory of the implicate order', Sheldrake's 'theory of formative causation', Lovelock's Gaia hypothesis, Prigogine's 'theory of dissipative structures', Lorenz and Feigenbaum's 'chaos theory' and Bell's theorem of non-locality, all of which insist that the scientist be capable of emphatic identification with the object he or she seeks to understand.[30] Because human understanding is not unequivocally compelled by the data to adopt one metaphysical position over another, a large element of human choice now supervenes. Into the epistemological equation there now enter more open-ended factors such as will, imagination, hope and empathy – all of which derive from what makes us human.

So too we have also transformed our relationship to our own bodies as medical science has rolled back the frontiers of death, or, in Paul Virilio's words, extended our 'health horizon'. Most of the great moral and ethical questions of our age concern the future of the body. Customizing genetic changes into a child, either before conception or during foetal development, now exercises moralists. So too do new breakthroughs in reproductive technologies including the freezing of sperm, *in vitro* fertilization and embryo transplantation, all of which make possible the artificial manipulation of the unborn. The great ethical debates of the age revolve around whether we have the right to shape the genetic destiny of human beings before birth. And all, of course, involve existential questions about who we are as a species.[31]

Freed from the tyranny of metaphysics, we can now ask whether we can eliminate cruelty given that there is nothing beyond our human incarnation (including a soul, an entity largely debunked by psychology). It is an ethical question of the first order, one that also involves a sense of the sacred. But the sacred now inheres not in God but in our biological self. In the process we are redefining our humanity, or what it means to be human. The ethical questions that bioethics raise, including how far we should go with genetic engineering and what our responsibilities are to the unborn, are addressed to one another, not to God. We are the sole source of the moral commandments about what we should or shouldn't do.[32]

The question that raises, in turn, is whether by eliminating pain to the body, we might reduce it even in war to an absolute minimum. Should we aspire to go beyond the 'death barrier'?

Beyond the 'Death Barrier'

In a Los Alamos briefing paper from the 1990s there is talk of maximizing force while reducing its lethal consequences. Non-lethal weapons (NLW) are described as the weapons of the future, weapons that for the first time in history can 'degrade the functioning threat to material or personnel without crossing "the death barrier"'. The language is jargon at its worst but the intention is clear enough, to pioneer a new form of warfare, one that might be able to take death out of the equation.[33]

Is that the real aim, however, or should we look for other motives behind it? In trying to eliminate the cruelty of war, is the United States aspiring to be less cruel physically, but no less cruel morally? Is the ethical conundrum of perceived violence leading it to conceal the violence implicit in war as the will to power?

To answer that question we must turn to Michel Foucault's *Discipline and Punish*, in which he describes how attitudes to pain have changed over the centuries. Foucault himself was interested in modern penology and the question he asked was how the coercive, corporeal, solitary, secret model of the power to punish replaced the representative, scenic, signifying, public, collective model. He was writing about the shift towards the end of the eighteenth century from punishment as public spectacle to private confinement. Instead of making punishment as spectacular and painful as possible, the prisoner's body was removed from society, subjected to constant surveillance and turned into a quantifiable subject with no capacity for agency. This too was 'disenchanting', for in lifting pain from the body, it imposed it on the spirit. True to the industrial principles of the time, penal reformers set out to 'redesign' the criminal by way of rehabilitating him and reintroducing him into society.[34]

This is not the first example of changes in cultural attitudes to punishment. Another is the great change halfway through the Middle Ages when the Fourth Lateran Council (1215) outlawed trial by ordeal. Deeming it primitive to invoke God's judgement, it preferred to invoke the justice of men. The Church resolved that man, not God, should determine guilt and looked forward to the high-tech forensic equipment then coming on the market: the thumbscrew, pincers and the rack. Smart penal reform now advocated torture.

The main value of Foucault's study is its probing of how attitudes change significantly over time. The history of mentalities pioneered by French historians in the early 1960s involves a stress on collective attitudes, not so much conscious thought as spoken or unconscious assumptions, as well as a concern with how people think, what they think, the structure of beliefs and their content. These historians offer us what Nietzsche first proposed – a historical anthropology of ideas.

We have never been more urgently in need of a historical anthropology of war than we are at present. Only through producing one can we really spot the *Zeitgeist*, or see where it is taking us. 'We are practising the non-lethal use of force', claimed a British general in Iraq, discussing the peacekeeping methods of the British Army in Basra and beyond. This is not the same, of course, as the use of non-lethal weapons.[35] Now a growing industry, books on NLW have been appearing for some time, yet very few have asked the essential questions about how their use – indeed, their very conception – indicates a specific and dramatic change in our attitudes to war.

One question they raise is a cynical one: has western society created a plethora of oxymoronic products or experiences designed to be precisely what they are not? Can the West only be cruel by changing the terms of engagement? To quote Slavoj Žižek: 'on today's market, we find a whole series of products deprived of their malignant properties: coffee without caffeine, cream without fat, beer without alcohol. . . . And the list goes on: . . . the Colin Powell

Doctrine of war with no casualties (on our side, of course), as warfare without warfare'.[36] But are we now aiming for a casualty-free experience for our enemies as well? Is this 'war without warfare', as Žižek claims?

It is all becoming very confusing. The extent of that confusion is nicely illustrated by the story of a US Navy SEAL sniper in Somalia in 1995 who was ordered to detune his special 'dazzling laser' to prevent him from blinding Somali citizens. Instead, he was ordered to shoot anyone who threatened him. 'We're not allowed to disable these guys because that was considered inhumane?' asked the bemused soldier. 'Putting a bullet in their head is somehow more humane?'[37] If we are trying to reinvent war, or trying to humanize it, it is not surprising that such questions abound.

For the militaries of the western world are developing a range of new weaponry, from acoustic weapons that shatter windows and cause internal damage to an enemy soldier, to electromagnetic pulse beams designed to knock down individuals and cause seizures, to chemical agents that can act as calmatives. The military will soon be able to take its pick from a range of kinetic, mechanical, chemical, electroshock, biological, acoustic, electromagnetic pulse, microwave and laser weapons, most of them second generation for they have not yet materialized as working technology but almost certainly will by 2030. Which is why the future of war lies in the future.

Does this mean that we have gone beyond the ethics of war as traditionally understood? Not necessarily. To really understand what is happening we have to go back to Foucault, who argued that the object of the penal reforms of the late eighteenth century which reduced modes of punishment, often of unsurpassable barbarity, to the relatively humane concept of incarceration was not to 'punish less', but to 'punish better'. The real purpose was to insert the power to punish more deeply into the body social: 'to constitute a new economy and a new technology for the power to punish'.

Clearly, as the technologies of non-lethal weapons are purely western and are likely to be used almost exclusively against non-western societies, their use can be seen as one means by which the

West cannot punish less but punish better with a clearer conscience. Yet Foucault was once asked, 'is it horrible to recognize that there are degrees of horror? Does recognizing that . . . there can be a humanization of its modes of existence mean defending prisons?'[38] He evaded the question. Is it horrible, however, to recognize that there can be degrees of horror in war too? Does recognizing that we can humanize its modes of existence mean that we are necessarily unethical in our concerns? Or do NLW promise to bring us closer to recognizing that if war will continue to remain central to the human condition (as too, presumably, will penal incarceration), we are ethically required to lessen its pain for its participants, willing and unwilling alike.

5

The Death of Sacrifice

> The computer chip may very well be a most useful war fighting tool. For example, while it is never a good thing when we lose a Predator on the battlefield, given the alternatives I look forward to many more computer chips dying for our country.
>
> US Secretary of the Air Force

> Any sufficiently advanced technology is indistinguishable from magic.
>
> Arthur C. Clarke

The metaphysical dimension of war, however, involves two ethical debates, not one. The first is the ethics of killing the enemy (the ethics of sacrificing him for a common good). As I have explained in the last chapter, there is nothing intrinsic in the future of war, such as speed and distance (both physical and real), that suggests either that its practice will be unethical, or that we need to produce a new ethical discourse for a new age.

But ethics also encompasses sacrifice. Without sacrifice, war has always been considered morally questionable. Few warriors have derived any public satisfaction from killing non-belligerents (those who, unarmed themselves, offer no threat to the armed). Without sacrifice or the willingness to risk death oneself, war cannot be 'sacred', and if it cannot be that it cannot be ethical. The words 'sacred' and 'sacrifice' are intimately linked.

But recent developments must prompt us to ask whether the United States, in instrumentalizing war once again, is not also likely to remove the concept of sacrifice from the equation. It is a question that can be asked but not answered conclusively because the technologies, especially robotics, that might allow it to aspire to that aim are still in their infancy. For want of a better word, they are 'magical' – and there is no more 'magical' image, as we can see from the popularity of films like *Terminator*, than the coming of the robot. We must conclude, then, where we began – with magic. It was the demise of magic that explained – for Max Weber – the disenchantment of the world. But could the reintroduction of magic (or what passes for it today) also be a source of future disenchantment?

The Greeks were the first to spawn myths of robots: human-like creatures made from metal. Hephaestus, who forged Achilles's shield, was said to have also created and brought to life a bronze statue for Minos of Crete, an archetypal robot that protected the island from attack. In similar fashion the bronze statue of Talos was created by Daedalus and animated to guard the sacred island of Thera, where the Argonauts came across it, much to their distress. The robots of the historical imagination reflect the possibilities of their age.

Thus to a common soldier, glimpsing an aerial dogfight in World War II, the future seemed clear enough:

> And that there were men in these expensive machines . . . was unimportant – except for the fact that they were needed to manipulate the machines. The very idea itself and what it implied . . . his powerlessness. . . . It was terrifying.[1]

In Nash's Battle of Britain pictures the machine, not the pilot, is triumphant; the machine is dominant in a vacuum, the sky. Nash could still evoke beauty: the crashing planes are all vertical wings and spirals, a play of intertorn vapour trails above a cubistically flattened landscape. Yet as Peter Conrad writes, it takes an effort to remind ourselves that the loops of black smoke record the track of a downed plane plummeting into the water, and an even greater effort to remember that the planes are not autonomous instruments of war but piloted by men.[2] Above all, there is no indication as to which side the plane belongs. In the abstract expanse of air, does it really matter?

Well, our present robots still take sides; they are programmed to do so. Sixty years after World War II, a Predator drone commanded by three operators on the ground in a simulator booth 600 miles away discharged two hell-fire missiles at a speeding sports utility vehicle in Yemen, killing its six terrorist occupants. A contemporary report stated that 'the robotic Star Wars style operation introduced a new weapon in the war against terror'.[3] The real innovation was a new, ultra-minimalist operation that fully integrated man, machine and software into 'a thinking system'. And indeed, the pilotless drones which were used on a large scale in the first Gulf War were the first real robots. A high-level study for the US Army, *Star 21: Strategic Technologies for the Army of the Twenty-first Century*, concluded that while the core twentieth-century weapon system had been the tank, in the twenty-first century it would be 'the unmanned system'. The study predicted that robots would be 'running and walking' by 2020.[4]

Obviously, robots have their uses. Drones already reconnoitre the battlefield. One day there will be robots that will be able to gather intelligence on the ground, perform sentry duty, prepare equipment and, as they do already, defuse bombs. The prime function of military robots will be to replace humans in particularly dangerous or tedious functions such as mine clearance, or in environments where biological or chemical weapons might be used. The real breakthrough will come when robots have combat use, particularly in land warfare, which poses a far more challenging operating environment for autonomous systems than does third-dimension combat.

The decisive point will be reached when robots, computer operated to collate data and synthesize new concepts out of previously obtained information, have to make value judgements (which is how decisions are derived). The United States has already developed a prototype combat aircraft, LOCAAS, as well as MANTA, an underwater vehicle with basic intelligent capabilities. To give systems intelligent capabilities such as planning and reasoning, it will have to go much further – to introduce parallel processing. At the moment, robots are engaged in sequential programming: processing data step by step.[5] Human beings, by contrast, can process different sequences at once, which allows real-time thoughts, memories, sensations and deductions. On the day that robots can engage in parallel processing, the nature of war really will have changed, instrumentally and existentially as well as metaphysically, for their use in war will render obsolete the concept of sacrifice.

The challenges that robots on the battlefield pose are many. One is whether we should allow the humanity of our species to be displaced by the needs of efficiency and utility. We have long held that success or failure in war should, at least, be distantly related to our own virtues: courage, endurance, perseverance, all three of which are deeply rooted in human nature. A more accurate claim would be that success or failure is rooted in the difference between our ideals and reality, and our capacity as moral beings to narrow the gap between the two. The question we would pose if we were ever to programme robots to kill is whether we should allow an autonomous battlefield system to search out enemies, redefined as a software algorithm.

If they see others in such terms, will they see us in that way too? Our fear of robots in this respect has a long history. We find it in a film script written by Roman Rolland in 1921 called *The Revolt of the Machines*. In this unrealized film, the machines revolt against their human masters when they send in tanks against other machines. The tanks change sides and take their crews prisoner. In Capek's play *R.U.R.*, humans make the mistake of turning robots into soldiers in the Balkans who promptly go on the rampage. In defining humanity as the problem, millions are killed. It is a logical 'end' to the knowledge-is-power ethos which began with the scientific revolution, for it is the firearm itself rather than simply the fire which opens up to the robots the ability to realize their own promise or potential.[6]

Secondly, should we allow ourselves to delimit the scope for human responsibility on the battlefield of the future? Will the machines eventually take over? Hans Moravec, a great guru of our robot future, predicts that by mid-century robots will take over the economy. Human beings, by exercising purchasing power, might still determine which robot companies returned a profit or loss, but they won't be running the companies any more. By analogy, in war we could still determine whether to fight or not, but once war is underway we'll have little influence over its conduct. What the robot/machine might do is something precise: it won't make the rules of war but it might, as Gödel's theorem implies, keep inventing even greater and more complex problems in an attempt to subject war to mathematical axioms. Machines may want to change the nature of war as machine science comes to resemble human science even less than quantum science resembles the physics of Aristotle.[7] And they may remain as competitive as man. For knowledge is never an end in itself. Most organisms, silicon and carbon, are compelled to seek knowledge that helps them survive the immediate future.

There is an ethical dilemma as well which takes us back to Asimov's Laws of Robotics, which he formulated with John Campbell and which are still the main litmus test of what we feel should be the operational parameters of robot action:

1. A robot may not injure a human being or, through inaction, allow a human being to come to harm.
2. A robot must obey the orders given it by human beings, except where such orders would conflict with the First Law.
3. A robot must protect its own existence as long as such protection does not conflict with the First or Second Laws.[8]

The question here is, what is a human being? A reasonable man or woman? Or a man or woman the robots may wish to disarm, whom they may wish to see reason?

Asimov's robots, of course, are supremely *ethical* machines. The ethical is encoded (or programmed) into them. They are programmed, after all, to save human life. The ethical imperative is central to their conception. To programme a machine to kill, by comparison, is to create a form of intelligence that is inhuman. In programming robots to kill, one denies them any choice. In killing they suffer no guilt, conscience or mortification. Because they have no rooted memory of ethical codes, taboos or conventions learned outside military culture (especially those learned in childhood), they have no tradition by which to judge their own actions.

Elaine Scarry puts the point very well in *The Body in Pain*. For we have already reached a defining moment in war:

> So completely have the formerly embodied skills of weapon use been appropriated into the interior of the weapon itself that no human *skill* is now required; and because the need for human skills is eliminated, the need for human *presence* to fire it is eliminated; and because the human presence is eliminated, the human act of *consent* is eliminated. The *building in of skill* thus becomes in its most triumphant form, the *building out of consent*. It is, of course, only at the 'firing' end of the weapon that human presence is eliminated; their body's presence at the receiving end is still very much required.[9]

What she argues is that the 'skill' of the soldier is now *embodied* not in him but in his weapon system – so why not remove the human presence altogether if one can? In a war fought by robots there would be no consent because there would be no human interdependence. There would be no sense of belonging to the same world. The machine would replace war with violence: an amoral indifference to human life defined as the relationship we have with one another. Even seeing the enemy as human would become optional, a matter of computer programming.

If that is the instrumental challenge of robotics in war, there is also an existential one. The real problem of robots can be found in Arendt's *The Human Condition*. In robbing life of danger, risk, pain and effort, we also threaten to rob it of its vitality too. Pain and effort are not just symptoms that can be done away with without changing life itself – they are modes through which life makes itself felt. That's why for the ancient Greeks the easy life of the gods, though attractive on one level, would have been lifeless had they led it themselves. For life depends on the intensity with which it is felt, and intensity can be experienced only when we talk of a burden, toil or the trouble of living it to the full. This, writes Arendt, is the difference between trust in the reality of the world and the reality of life. For although the world endures without us, life *is* us. We either live it to the full or we do not, and what we get out of life is what we invest in it.[10]

Here we must draw a distinction between tools and robots, one for which we must go back to Aristotle. He drew a distinction between an instrument designed for action and one designed for production. An instrument of production was one which produced an 'effect'. An instrument of action was one from which one gained nothing but its own use, one which was 'used' by men of action. What makes a man 'act' is when he does things, not when he makes them. 'Life is action, not production', insisted Aristotle.[11]

Technology in the last 2,000 years has certainly enhanced action, or our sense of agency. But it has not replaced it – so far. The tools or instruments that ease (or enhance) our labours are products of our work. They free us from necessity. Once we eliminate the need for work – once we free ourselves altogether from necessity – then we cease to be free agents. For our freedom comes entirely from our struggle against necessity. In that sense replacing warriors with robots would rob war of its existential appeal.

It would also deny it metaphysical content as well. It is to be expected that ethical problems arise when communities have different preferences for ways of living, preferences which we can see as culture specific. All societies choose the best life for them, in the light of their own historical experience and collective self-understanding. It is at this point that ethical differences between communities arise.

But we also abide by universal codes, by the demands of human rights, which touch on an altogether different question, what the contemporary philosopher Jürgen Habermas calls our 'self-understanding as members of the same species'. This concerns not culture, which is different in every age and every society, but the vision that different cultures have of humanity, which is the same. Our recent sacralization of the human heart as well as the human body stems from this recognition, one which these days is largely grounded in biology: the recognition that we all feel pain and should do what we can to avoid it or, even better, eliminate it altogether.

For Habermas, the biotechnological revolution threatens this self-understanding of the species. Indeed, he believes that recent developments in biotechnology and genetic research threaten to instrumentalize human nature according to instrumental preferences. The most obvious example is parents who want children of a certain skin colour, or hair colour, or parents who would like to breed out what they consider human imperfections, most of them genetic. The human body at this point is no longer sacred because it becomes an 'object' or an instrument of parents or the state, to be modified or redesigned at will. And it is at this point, for Habermas, that our self-understanding as a species may be threatened. We may see human bodies merely as defective 'hardware' and the mind as enhanced 'software'.[12]

Habermas has been criticized for his pessimism, but his fears about technology could be applied, perhaps with greater force, to robotics. For robots programmed to kill other human beings threaten to deny us our humanity in a much more direct and immediate fashion. We have always taken the ultimate decision who to kill or not to kill; who to take hostage or pardon or sell into slavery; who to take prisoner and how to treat the prisoners we have taken. At times we have been ruthless in our judgements, especially in the medieval period when townsfolk who refused to yield to besieging armies automatically forfeited the right to be treated fairly when captured. Power is what we are willing to give up, not only what we enjoy – to give up the right to slay a defeated enemy and to show mercy instead is the application of power at its most unconditional, and we show mercy when we accept 'species being' rather than identify the defeated enemy – whether a nation, class or social group – as 'subhuman', or 'quasi-human', or 'not quite human'.

If robots are to be true to Asimov's first principle – that they should not harm humans – then it seems to me that the only way we can programme them to kill is by persuading them (through human programming) to accept their targets (other human beings) as being 'objectively non-human'. Some of the research into nanotechnology and robotics in the research laboratories of the United States offers a chilling future – for what could be more disenchanting than a future in which we allow autonomous machines to take the decisions for us, in which we allow robotic systems to target us as 'carbo units' rather than creatures of flesh and blood?[13]

6

To Be Concluded?

The way in which war is conducted today can seem disenchanting. In an article entitled 'Was There a Gulf War?' (a title which echoed, consciously or otherwise, Baudrillard's famous article 'The Gulf War Did Not Take Place'), the science fiction writer J. G. Ballard dismissed the conflict as surreal – an arcade video game of a bombing campaign, followed by a hundred hours of ground fighting filtered through military and TV censors. To Ballard the final image of the campaign, the devastation on the Basra highway, looked like 'a traffic jam left out to rust, a discarded Mad Max film set, the ultimate autogeddon'. In the absence of any pictures of the dead or wounded, it created the impression that the entire war was a vast demolition derby.

Earlier, in a review of one of the *Mad Max* films, Ballard had written that it afforded a compellingly reductive vision of post-industrial collapse, as gangs of motorized savages roved the desert wastes bereft of speech, thought, hopes or dreams, dedicated only to the brutal realities of speed and violence.[1] Speed and violence – the twin features of post-industrial warfare conjured up by another French writer, Paul Virilio; or, to be more accurate, speed in the service of violence, the organized violence of postmodern war.

Ballard echoes many of the familiar clichés about the way we do war – media-mediated conflict, long distance both geographically and emotionally, disembodied physically and figuratively. Is he right?

I have tried to argue throughout that this is a caricature, an intellectual distortion. What I have set out to argue in this short book is that war has a future. That future, as with so much else in the world, lies with biotechnology, a revolution that is already changing many of our assumptions about politics, social transformation and human behaviour.

It would now seem, as evolutionary psychologists are telling us, that human behaviour is far more genetically determined than we thought before, and that by enhancing, modifying or altering our genes, we may well be able to enhance the things we do well, and have always done well, as a species. One of the things we have done particularly well over the centuries has been war, and there is nothing to suggest that we will be going out of the war business – indeed, quite the opposite.

The 'end of war' first dreamed of by the Kantians, and subsequently by other writers of a cosmopolitan bent, has a history – but not a very long one. It was in essence an Enlightenment belief that human nature could be changed through the environment, that humans could be educated to be better citizens, or less wilful neighbours. If the environment was changed through political action – Kant's republic, or Marx's socialist utopia – conflict could be programmed out of the human species.

But from the perspective of the early twenty-first century, nature is being privileged over nurture as never before. The importance of genetics in shaping human behaviour and personality is now too widely grounded in evidence to deny. Add to this experiments in the neurosciences which reveal how human perceptions and behaviour are produced and determined by physical processes within the brain (another instance where the information age helps us understand our own biology, and vice versa), we must acknowledge that all behaviour is cross-cultural, and thus not determined primarily by the environment but by a mixture of genetic and environmental factors.

The human genome is a universal one, and what its decoding tells us is that genetics determines behaviour more than we once thought. None of which is to say that we should engage in genetic determinism, any more than in the political determinism of the past. Cultures evolve norms to channel human aggression into creative outlets. Cultures both prescribe and proscribe human behaviour very effectively. But war is probably too genetically grounded to eradicate easily. For it stems from the same genetic mix that allows us to entertain the idea of peace. As we know from coalition-forming in chimpanzee colonies, war and peace have the same origins: the ability to cooperate. To cooperate means the chance to build alliances for peaceful or violent ends.

Developed societies (the sole focus of this study) are likely, therefore, to continue with war but in a form that is more rational and optimal than ever before. By exploiting technological changes, many of them associated with the information revolution, we may be able to humanize war, or to make it more humane. We may even be able to pass beyond the 'death barrier'. For if war is seen as merely one end on a spectrum of violence, death is not necessarily essential to it. Killing could be made redundant (though probably not optional), leaving physical coercion or the will to power by other means. Dying could be minimized as well, while still ensuring that the warrior persona remains in some shape or form, genetically if not mimetically. It is the future of sacrifice – the metaphysical or third dimension of war – that is the most difficult to omit from the equation without transforming the nature of war. And if we do that, we may run the risk of making it far more 'disenchanting' than before. Indeed, it is not impossible to imagine an age in which war could be far more soulless than ever, even if fewer people get killed. A cold, rational war machine could well evolve, one that could be considered 'inhumane' in different ways.

This, it seems to me, is the great challenge. In a secular world, writes Steven Lukes, our commitment to sacred values (including sacrifice) is much more conditional than it was. Would it not be better to get rid of the sacred values altogether; to carry further what Max Weber called 'the disenchantment of the world', to follow the advice of the great nuclear theologian Tom Schelling and subject all choices (including the saving of lives) to a cost–benefit analysis? Would it not be better to see war in terms of trade-offs rather than sacrifices, as would befit an entirely instrumental practice?

We have a choice. There is an instrumentalist vision in the way many Americans look at war, one which has a long history – the 'technicization' of the life world (including war) through instrumental reason. There is another vision we associate with the suicide bomber and postmodern terrorist: the sacralization of war, the privileging of the spirit (and the attendant triumph of the will) over the body. Both are extremes, and as Lukes warns, we should fear a world in which either has won out.[2] We should aim instead to position ourselves in between – to make war valuable morally is to make it sacred in terms of sacrifice, the willingness to die for a cause in which one believes. To take out sacrifice would be to disenchant it once again. To surrender responsibility to computers with artificial intelligence who would feel no guilt, and no remorse for their actions, or to autonomous weapons systems with no concept of loss, would be to compromise the human dimension of war.

It is a world, however, which many scientists are trying to forge. But that, as they say, is the subject of another book . . .

Notes

Preface

1 Kurt Vonnegut, *Slaughterhouse-Five* (London: Vintage, 1991), pp. 52–3.
2 Charles Jonscher, *Wired Life: Who We Are in the Digital Age* (New York: Anchor, 1999), pp. 258–9.

1 The Re-enchantment of War

1 A. N. Wilson, *The Victorians* (London: Hutchinson, 2002), p. 138.
2 Cited in Mark Rose, *Alien Encounters: An Anatomy of Science Fiction* (Cambridge, MA: Harvard University Press, 1981), p. 141.
3 Cited in Wilson, *The Victorians*, p. 144.
4 Cited in Daniel Pick, *The War Machine: The Rationalization of Slaughter in the Modern Age* (New Haven, CT: Yale University Press, 1993), p. 72.
5 Cited in Ian Ousby, *The Road to Verdun* (London: Jonathan Cape, 2002), p. 67.
6 H. H. Gerth and C. Wright Mills, *From Max Weber: Essays in Sociology* (London: Routledge, 1997), p. 155.
7 Peter Lassman, 'The Rule of Man over Man: Politics, Power and Legitimation', in Stephen Turner (ed.), *The Cambridge Companion to Weber* (Cambridge: Cambridge University Press, 2000), p. 97. Not that the idea of 'disenchantment' is new. The revolution in thought associated with the mechanistic thinking of philosophers such as Descartes made some Europeans aware of the intellectual world they had lost. The eighteenth-century scholar Richard Hurd wrote, 'What we have gotten by this revolution, you will say, is a great deal of good sense. What we have lost is a world of fine fabling' (cited in Peter Burke, *Varieties of Cultural History* (Cambridge: Polity Press, 1997), p. 14).

8 Cited in Paul Kerr, *The Crimean War* (London: Boxtree, 1997), p. 178.
9 Cited in Richard Schlacht, 'The Future of Human Nature', in Paul Gifford (ed.), *2000 Years and Beyond: Faith, Identity and the 'Common Era'* (London: Routledge, 2003), p. 77.
10 Marcel Gauchet, *The Disenchantment of the World: A Political History of Religion* (Princeton, NJ: Princeton University Press, 1988), p. 34.
11 Cited in Theodore Schieder, 'The Role of Historical Consciousness in Political Action', *History and Theory*, 27, 4 (1978), p. 2.
12 Cited in Ousby, *Verdun*, p. 23.
13 Curzio Malaparte, *Kaputt* (1944; London: Picador, 1982), p. 161.
14 For a general discussion of Hegel and war see my *War Without Warriors: The Changing Culture of Military Conflict* (New York: Lynne Rienner, 2002), pp. 54–7.
15 Cited in Harold Bloom, *Genius: A Mosaic of a Hundred Exemplary Creative Minds* (London: Fourth Estate, 2002), p. 506.
16 Cited in George Steiner, *Errata: An Examined Life* (London: Weidenfeld and Nicolson, 1997), p. 16.
17 See my discussion of Ernst Junger in *War in the Twentieth Century: The Impact of War on Modern Consciousness* (London: Brasseys, 1994), pp. 118–25.
18 Michael Evans, 'Close Combat: Lessons from the Cases of Albert Jacka and Audie Murphy', in Michael Evans and Alan Ryan (eds), *The Human Face of Warfare: Killing, Fear and Chaos in Battle* (London: Alan and Unwin, 2000), p. 44.
19 Ernst Junger, *Storm of Steel*, trans. Michael Hoffmann (London: Allen Lane, 2003), p. xix.
20 Ernst Junger, *Storm of Steel* (London: Constable, 1994), pp. 307–8.
21 Christopher Logue, *All Day Permanent Red: The First Battle Scenes of Homer's Iliad* (New York: Farrar, Straus and Giroux, 2003).
22 Simone Weil, *The Iliad, or the Poem of Force*, trans. Mary McCarthy (Pendal Hill: Pamphlets, 1957).
23 Susie Giblik, *Has Modernism Failed?* (London: Thames and Hudson, 1984), p. 93.
24 Roger Scruton, 'Desecrating Wagner', *Prospect* (April 2003), p. 40.
25 Cited in Stefan Elbe, *Europe: A Nietzschean Perspective* (London: Routledge, 2003), p. 47.
26 Cited in Louis Dumont, *German Ideology: From France to Germany and Back* (Chicago: Chicago University Press, 1994), p. 231.
27 Hannah Arendt, *The Human Condition* (Chicago: Chicago University Press, 1998), p. 181.
28 Martyn Lyon, *Napoleon Buonaparte and the Legacy of the French Revolution* (London: Macmillan, 1994), p. 46.

29 Samuel Hynes, *The Soldier's Tale: Bearing Witness to Modern War* (London: Pimlico, 1998), p. 17.
30 Michael Handel, *Masters of War: Classical Strategic Thought* (London: Frank Cass, 1992), p. 5.
31 Cited in A. D. Harvey, *Collision of Empires: Britain in Three World Wars, 1793–1945* (London: Phoenix, 1992), p. 51.
32 Cited in David Chandler, *The Campaigns of Napoleon* (London: Weidenfeld and Nicolson, 1994), p. 179.
33 Cited in Rory Muir, *Tactics and the Experience of Battle in the Age of Napoleon* (New Haven, CT: Yale University Press 1998), p. 42.
34 Cited in J. McManners, *Lectures in European History, 1789–1914* (Oxford: Blackwell, 1966), p. 34.
35 Winston Churchill, *The World Crisis 1911–18*, abridged and revised (London: Macmillan, 1943), pp. 649–50.
36 Julian Young, *Heidegger's Later Philosophy* (New York: Cambridge University Press, 2001), p. 48.
37 Nicholas Humphrey, *Consciousness Regained: Chapters in the Development of the Mind* (Oxford: Oxford University Press, 1984), p. 18.
38 H. A. Simon, *The Sciences of the Artificial* (Cambridge, MA: Massachusetts Institute of Technology, 1969), p. xi.
39 Cited in Nathan Schlanger, 'Mindful Technology: Unleashing the 'Chaine Opératoire' for an Archaeology of Mind', in Colin Renfrew and Ezra Zubrow (eds), *The Ancient Mind: Elements of Cognitive Archaeology* (Cambridge: Cambridge University Press, 1994), p. 149.
40 Edith Wyschogrod, *Spirit in Ashes: Hegel, Heidegger and Man-made Death* (New Haven, CT: Yale University Press, 1985), p. 189.
41 Young, *Heidegger's Later Philosophy*, p. 103.
42 Frederic Ferre, *Philosophy of Technology* (Athens: University of Georgia Press, 1995), p. 65.
43 Ibid.
44 Cited in Pick, *The War Machine*, p. 53.
45 Agnes Heller, *A Theory of Modernity* (Oxford: Blackwell, 1999), p. 164.
46 Peter Conrad, *Modern Times, Modern Places: Life and Art in the Twentieth Century* (London: Thames and Hudson, 1998), pp. 253–54.
47 Paul Fussel, *Wartime: Understanding and Behaviour in the Second World War* (Oxford: Oxford University Press, 1989), p. 273.
48 Hynes, *The Soldier's Tale*, p. 58.
49 Cited in James Gleick, *Faster: The Acceleration of Just About Everything* (London: Abacus, 1991), p. 35.
50 Bernard Doray, *From Taylorism to Fordism: A Rational Madness* (London: Free Association, 1988), p. 49.

51 Robert Kanigel, *The One Best Way: Frank Winslow Taylor and the Enigma of Efficiency* (New York: Viking, 1997), pp. 339–40.
52 Fernand Celine, *Journey to the End of Night* (New York: Giroux, 1983), p. 99.
53 Cited in Enzo Traverso, *Origins of Nazi Violence* (New York: New Press, 2003), p. 81.
54 Heller, *Theory of Modernity*, p. 107.
55 Cited in Conrad, *Modern Times, Modern Places*, p. 401.
56 Manuel de Landa, *War in the Age of the Intelligent Machine* (New York: Zone, 1991), p. 66.
57 Bernard Lewis, *What Went Wrong With Islam? The Clash Between Islam and Modernity in the Middle East* (London: Weidenfeld and Nicolson, 2002), p. 124.
58 Werner Heisenberg, *Physics and Technology* (London: Penguin, 1990), p. 51.
59 George L. Mosse, *Fallen Soldiers: Reshaping the Memory of the World Wars* (Oxford: Oxford University Press, 1990), p. 61.
60 Paul Virilio, *Speed and Politics* (New York: Semiotext(e): Foreign Agents Services Autonomeda, 1986), p. 23.
61 Paul Johnson, *Napoleon* (London: Weidenfeld and Nicolson, 2002), p. 32.
62 Jeremy Rifkin, *The Biotech Century: How Genetic Commerce Will Change the World* (London: Orion, 1998), p. 26.
63 Ibid.
64 Ibid., p. 209.
65 Tom Siegfried, *The Bit and the Pendulum: From Quantum Computing to M Theory: The New Physics of Information* (New York: John Wiley, 2000), p. 101.
66 Stephen Toulmin, 'From Clocks to Chaos: Humanizing the Mechanistic World View', in Hermann Haken, Anders Karlqvist and Uno Svedin (eds), *The Machine as a Metaphor and Tool* (Berlin: Springer, 1993), p. 152.
67 Rifkin, *Biotech Century*, pp. 182–3.
68 Ibid., p. 184.
69 Cited in Steven Metz, *Armed Conflict in the Twenty-first Century: The Information Revolution and Post-modern Warfare* (US Army War College: Strategic Studies Institute, April 2000), p. 30.
70 Fussel, *Wartime*, p. 9.
71 Charles Jonscher, *Wired Life: Who We Are in the Digital Age* (New York: Anchor, 1999), p. 120.
72 Ibid.
73 Fussel, *Wartime*, p. 14.
74 Stig Dagerman, *German Autumn* (London: Quartet, 1988), p. 59.
75 Ferre, *Philosophy of Technology*, p. 134.
76 Cited in Alister McGrath, *The Re-enchantment of Nature: Science, Religion and the Human Sense of Wonder* (London: Hodder and Stoughton, 2002), p. 95.

77 Lewis Mumford, *Art and Technics* (New York: Columbia University Press, 2000), p. 78.
78 Ferre, *Philosophy of Technology*, p. 134.
79 Cited in Metz, *Armed Conflict in the Twenty-first Century*, p. 34.
80 Ibid., p. 28.
81 H. G. Wells, 'Changes in the Arts of War', in H. G. Wells, *The Way of the World: Guesses and Forecasts of the Years Ahead* (London: Enos-Benn, 1928), p. 146.

2 The Warrior of the Future: Memes or Genes?

1 Zygmunt Bauman, *Liquid Love: On the Frailty of Human Bonds* (Cambridge: Polity, 2003), p. 9.
2 Peter Ackroyd, *Albion: The Origins of the English Imagination* (London: Chatto and Windus, 2002), p. 119.
3 Ibid.
4 For alienation see Daniel Bell, *The End of Ideology: On the Exhaustion of Political Ideas in the Fifties* (Cambridge, MA: Harvard University Press, 2000), pp. 358–64.
5 Cited in Chris Baldick, *In Frankenstein's Shadow: Myth, Monstrosity and Nineteenth-century Writing* (Oxford: Clarendon Press, 1987).
6 Stephen Crane, *The Red Badge of Courage* (New York: Norton, 1962), p. 161.
7 Paul Fussel, *Wartime: Understanding and Behaviour in the Second World War* (Oxford: Oxford University Press, 1989), p. 66.
8 See Susan Blackmore, *The Meme Machine* (Oxford: Oxford University Press, 1999), p. 30.
9 Omer Bartov, *Murder in Our Midst: The Holocaust, Industrial Killing and Representation* (Oxford: Oxford University Press, 1996), p. 17.
10 Maria Michela Sassi, *The Science of Man in Ancient Greece* (Chicago: Chicago University Press, 2001), pp. 36–9.
11 I have derived this formula from Luc Ferry, *Homo Aestheticus: The Invention of Taste in the Modern Age* (Chicago: Chicago University Press, 1993), pp. 251ff.
12 Hannah Arendt, 'The Concept of History: Ancient and Modern', in Hannah Arendt (ed.), *Between Past and Future* (New York: Viking, 1961), p. 43. Arendt writes also of life being a stage on which one enacts an idealized version of one's own life. It is a stage on which one acts out what she calls the 'latent self', a phrase she derives from Dante. 'For in every action what is primarily intended by the doer . . . is the disclosure of his own image. Hence . . . every doer in so far as he takes delight in doing; since everything that is

desires its own being and since in action the being of the doer is somehow intensified, delight necessarily follows. . . . Thus, nothing acts unless [by acting] it makes patent its latent self.' In another book Arendt talks about this at length as the 'urge towards self-display'. See Peter Euben, *Platonic Noise* (Princeton, NJ: Princeton University Press, 2003), p. 55.

13 R. P. Winnington-Ingram, *Sophocles: An Interpretation* (Cambridge: Cambridge University Press, 1980), pp. 55–6.

14 John A. Lynn, *Battle: A History of Combat and Culture* (Boulder: CO: Westview, 2003), p. 6.

15 Jean-Pierre Vernant and Pierre Vidal-Naquet, *Myth and Tragedy in Ancient Greece* (New York: Columbia University Press, 1990), p. 62.

16 Martha Nussbaum, 'Tragedy and Self Suffering: Plato and Aristotle on Fear and Pity', in Julia Annas (ed.), *Oxford Studies in Ancient Philosophy* (Oxford: Clarendon Press, 1992), p. 27.

17 Adam Parry, 'The Two Voices of Virgil's Aeneid', in Steele Commanger (ed.), *Virgil: A Collection of Critical Essays* (Englewood Cliffs, NJ: Prentice-Hall, 1966), p. 114.

18 Herbert J. Muller, *The Uses of the Past: Profiles of Former Societies* (New York: Oxford University Press, 1966), p. 138.

19 E. R. Dodds, *Pagan and Christian in an Age of Anxiety* (Cambridge: Cambridge University Press, 1965), p. 8.

20 Marcel Gauchet, *The Disenchantment of the World: A Political History of Religion* (Princeton, NJ: Princeton University Press, 1988), pp. 128–9.

21 Cited in Philippe Contamine, *War in the Middle Ages* (Oxford: Blackwell, 1984), pp. 253–4.

22 Jonathan Riley-Smith, *The Crusades: A Short History* (New Haven, CT: Yale University Press, 1987), pp. 256–7. See also his 'Crusading as an Act of Love', in Thomas F. Madden (ed.), *The Crusades: The Essential Readings* (Oxford: Blackwell, 2002), pp. 31–51.

23 Riley-Smith, *The Crusades*, p. 132.

24 Norbert Elias, *The Civilizing Process* (Oxford: Blackwell, 2000), p. 165.

25 Riley-Smith, *The Crusades*, p. 49.

26 Harold Bloom, *Genius: A Mosaic of a Hundred Exemplary Creative Minds* (London: Fourth Estate, 2002), p. 34.

27 W. H. Auden, *Complete Works of W. H. Auden: Prose*, vol. 2 (1939–48), ed. Edward Mendelssohn (Princeton, NJ: Princeton University Press, 2002), pp. 365–6.

28 Agnes Heller, *A Theory of Modernity* (Oxford: Blackwell, 1999), p. 227.

29 Michel de Montaigne, *The Complete Essays*, ed. Michael Screetch (London: Allen Lane, 1991), p. 788.

30 Harold Bloom, *Shakespeare: The Invention of the Human* (London: Fourth Estate, 1999), p. 741. For Bloom one of the ultimate characters in this regard is that most noble and yet sinister of Shakespeare's heroes, Henry V. In the play, his most bombastic and patriotic, we find what Yeats calls 'an amiable monster'. His amiability, adds Bloom, lies in his self-knowledge. He is an honest hypocrite, a spinner of words. But there's darkness in his soul which we – and he – know is there. This comes out when he compares himself with Alexander the Great. Indeed, the Alexander motif runs through the play. Drunken Alexander murdered his good friend Cleitus. Shakespeare shows us a young king 'killing' his former friends on the campaign and breaking the heart of John Falstaff. We see him urging his men into the breach at Harfleur, extolling their fathers as 'so many Alexanders'. He has the glamour of an Alexander who staked all on one military enterprise and who, like his hero, dies young with other worlds still to conquer. But, adds Bloom, 'this is an Alexander endowed with inwardness . . . in Henry's vision the growing inner self requires an expanding kingdom and France is the designated realm for growth'.

31 John Hale, *War and Society in Renaissance Europe* (London: Fontana, 1985), p. 57.

32 Thomas Arnold, *The Renaissance at War* (London: Cassell, 2001), p. 87.

33 Ibid., p. 102.

34 Carl von Clausewitz, *Historical and Political Writings*, ed. Peter Paret and Daniel Morgan (Princeton, NJ: Princeton University Press, 1992), pp. 279–84.

35 George Steiner, *Grammars of Creation* (London: Faber and Faber), p. 274.

36 T. G. Rosenthal, 'War and the Artist', in A. J. P. Taylor and J. M. Roberts (eds), *Purnell History of the Twentieth Century*, vol. 2 (London: Purnell, 1968), p. 632.

37 Steiner, *Grammars of Creation*, p. 274.

38 Cited in A. D. Harvey, *Collision of Empires: Britain in Three World Wars, 1793–1945* (London: Phoenix, 1992), p. 349.

39 See my discussion in *War and the Twentieth Century: A Study of War and Modern Consciousness* (London: Brasseys, 1994), p. 119.

40 See Lawrence L. Langer, *The Age of Atrocity: Death in Modern Literature* (Boston: Beacon Press, 1978), pp. 69–112.

41 Cited in Domenico Losurdo, *Heidegger and the Ideology of War: Community, Death and the West*, trans. Marella Morris and Jon Morris (Amherst, NY: Humanity Books, 2001), p. 25.

42 See my *War and the Illiberal Conscience* (Boulder, CO: Westview, 1998), p. 65.

43 Julius Evola, *Men Among the Ruins* (Rochester, VT: Inner Traditions, 2002), p. 202.
44 James Blinn, *The Aardvark is Ready for War* (New York: Anchor, 1997), pp. 127–8.
45 Ibid., pp. 261–2.
46 Elaine Graham, *Representations of the Post-human* (Manchester: Manchester University Press, 2002), p. 184.
47 Ibid., p. 185.
48 David Tomas, 'Art, Psychasthenic Assimilation and the Cybernetic Automaton', in Chris Hables Gray (ed.), *The Cyborg Handbook* (London: Routledge, 1995), p. 260.
49 Cited in Graham, *Representations of the Post-human*, p. 121.
50 See the discussion by Shadia Drury, *Alexandre Kojève: The Root of Post-modern Politics* (London: Macmillan, 1994), pp. 185–6.
51 Gregory Stock, *Redesigning Humans: Choosing Our Children's Genes* (London: Profile, 2003), p. 13.
52 Ibid., p. 105.
53 Finn Bowring, *Science, Seeds and Cyborgs: Biotechnology and the Application of Life* (London: Verso, 2003), p. 212.
54 Jeremy Rifkin, *The Biotech Century: How Genetic Commerce Will Change the World* (London: Orion, 1998), p. 164.
55 David Tomas, 'The Technophilic Body: On Technicity', in Gray (ed.), *Cyborg Handbook*, p. 180.
56 Ibid.

3 Towards Post-human Warfare

1 Cited in Michael Llewelyn Smith, 'The War Poets', in A. J. P. Taylor and J. M. Roberts (eds), *Purnell History of the Twentieth Century*, vol. 2 (London: Purnell, 1968), p. 641.
2 Robert Kaplan, *Warrior Politics* (New York: Random House, 2002), p. 122.
3 Paul Fussel, *Wartime: Understanding and Behaviour in the Second World War* (Oxford: Oxford University Press, 1989), p. 68.
4 Cited in Adam Piette, *Imagination at War: British Fiction and Poetry, 1939–45* (London: Macmillan, 1999), p. 215.
5 Ibid., p. 214.
6 Ibid.
7 Ibid., p. 215.
8 Ibid., p. 214.

9 Samuel Hynes, *The Soldier's Tale: Bearing Witness to Modern War* (London: Pimlico, 1998), p. 125.
10 Chris Hables Gray, 'The Cyborg Soldier: The US Military and the Post-modern Warrior', in Les Levidow and Kevin Robins, *Cyborg Worlds: The Military Information Society* (London: Free Association Books, 1989), p. 47.
11 Cited in Piette, *Imagination at War*, p. 228.
12 Ibid., p. 231.
13 Cited in Chris Hables Gray (ed.), *The Cyborg Handbook* (London: Routledge, 1995), p. 322.
14 William Gibson, *Neuromancer* (London: HarperCollins, 1993), p. 37.
15 Joseph Heller, *Catch 22* (London: Vintage, 1994), p. 373.
16 Cited in Joseph Mazzeo, *Renaissance and Revolution: The Remaking of European Thought* (London: Methuen, 1967), p. 179.
17 Francis Spufford and Jenny Uglow, *Cultural Babbage: Technology, Time and Invention* (London: Faber and Faber, 1996), p. 270.
18 James Gleick, *Faster: The Acceleration of Just About Everything* (London: Abacus, 1991), p. 113.
19 Spufford and Uglow, *Cultural Babbage*, p. 277.
20 Gray (ed.), *Cyborg Handbook*, p. 23.
21 Charles Jonscher, *Wired Life: Who We Are in the Digital Age* (New York: Anchor, 1999), p. 128.
22 Hans Jonas, 'Cybernetic and Purpose: A Critique', in Hans Jonas (ed.), *The Phenomenon of Life: Towards a Philosophical Biology* (Evanston, IL: Northwestern University Press, 2001), p. 110.
23 Stephen Pinker, *The Blank Slate: The Modern Denial of Human Nature* (London: Allen Lane, 2002), p. 32.
24 Frederic Ferre, *Philosophy of Technology* (Athens: University of Georgia Press, 1995), p. 128.
25 Jonscher, *Wired Life*, p. 148.
26 Ibid.
27 Cited in Gray (ed.), *Cyborg Handbook*, p. 145.
28 Cited in Tim Jordan, *Cyberpower: The Culture and Politics of Cyberspace and the Internet* (London: Routledge, 1999), pp. 112–13.
29 Peter Conrad, *Modern Times, Modern Places: Life and Art in the Twentieth Century* (London: Thames and Hudson, 1998), p. 641.
30 David F. Channell, *The Vital Machine: A Study of Technology and Organic Life* (New York: Oxford University Press, 1991).
31 Bruce Mazlish, *The Fourth Discontinuity: The Co-evolution of Humans and Machines* (New Haven, CT: Yale University Press, 1993).
32 *The Times*, 16 October 2003.

33 Ray Kurtzweil, *The Age of Spiritual Machines* (London: Orion, 1999), p. 2.
34 Richard Rorty, *Philosophy and Social Hope* (London: Penguin, 1999), p. 52.
35 Adam Roberts, *Science Fiction* (London: Routledge, 2000), p. 8.
36 Orson Scott Card, *Ender's Game* (New York: Tom Doherty, 1991); Leo Frankowski, *A Boy and His Tank* (New York: Simon and Schuster, 1999).
37 Joanna Bourke, *An Intimate History of Killing: Face to Face Killing in the Twentieth Century* (London: Granta, 1991), p. 1.
38 Pinker, *Blank Slate*, p. 167.
39 Ibid.
40 A. D. Harvey, 'Soldiers with Operational Flair', *RUSI Journal* 147, 1 (February 2002), p. 61.
41 Richard Dawkins, 'What's Wrong with Cloning?', in Martha Nussbaum and Case R. Sunstein (eds), *Clones and Clones: Facts and Fantasies about Human Cloning* (New York: Norton, 1988), p. 55.
42 *The Times*, 19 November 2002.
43 David Grossman, *On Killing* (New York: Free Press, 1999), p. 270.
44 Francis Fukuyama, *Our Post-human Future: Consequences of the Biotechnology Revolution* (London: Profile Books, 2002), p. 52.
45 Michael Gelven, *War and Existence: A Philosophical Enquiry* (University Park, PA: Penn State University Press, 1994), p. 122.
46 *The Times*, 10 July 2003.

4 Beyond the Death Threshold

1 Friedrich Nietzsche, *The Gay Science*, ed. Bernard Williams (Cambridge: Cambridge University Press, 2001), p. 34.
2 Cited in Daniel Ahern, *Nietzsche as Cultural Physician* (University Park, PA: Penn State University Press, 1995), p. 31.
3 Ibid., p. 42.
4 Ibid.
5 Peter Paret, *Imagined Battles: Reflections of War in European Art* (Chapel Hill: University of North Carolina Press, 1997), pp. 70–6.
6 Cited in Michael Llewelyn Smith, 'The War Poets', in A. J. P. Taylor and J. M. Roberts (eds), *Purnell History of the Twentieth Century*, vol. 2 (London: Purnell, 1968), p. 643.
7 Cited in Hunt Tooley, *The Western Front: Battleground and Home Front in the First World War* (London: Palgrave Macmillan, 2003), p. 192.
8 Cited in Ian Ousby, *The Road to Verdun* (London: Jonathan Cape, 2002), p. 66.

9 Cited in Elmer Bendiner, *The Fall of Fortresses: A Personal Account of the Most Damaging and Deadly American Air Battles of World War II* (New York: Putnam, 1980), p. 138.
10 See Maria Tatar Lustmord, *Sexual Murder in Weimar Germany* (Princeton, NJ: Princeton University Press, 1995), p. 75.
11 Ernst Junger, *Storm of Steel*, trans. Michael Hoffmann (London: Allen Lane, 2003), p. 119.
12 Arthur C. Danto, *The Philosophical Disenfranchisement of Art* (New York: Columbia University Press, 1986), p. 103.
13 Milan Kundera, *Slowness* (New York: HarperCollins, 1996), p. 2.
14 Domenico Losurdo, *Heidegger and the Ideology of War: Community, Death and the West*, trans. Marella Morris and Jon Morris (Amherst, NY: Humanity Books, 2001), p. 173.
15 Mark Bowden, 'The Kabul-Ki Dance', *Atlantic Monthly* 290, 4 (2002).
16 Terry Eagleton, *The Ideology of Aesthetics* (Oxford: Blackwell, 1990), p. 1.
17 Jonathan Sacks, *The Dignity of Difference: How to Avoid the Clash of Civilizations* (London: Continuum, 2002), p. 32.
18 Susan Jeffords, *The Remasculization of America: Gender and the Vietnam War* (Bloomington: Indiana University Press, 1989), p. 14.
19 Walter Benjamin, 'The Story Teller', in *Illuminations* (London: Pimlico, 1999), p. 93.
20 John Ellis, *The Social History of the Machine Gun* (London: Pimlico, 1976), pp. 22–3.
21 James Surowiecki, 'Turn of the Century', *Wired* (January 2000), p. 85.
22 Robert Kanigel, *The One Best Way: Frederick Winslow Taylor and the Enigma of Efficiency* (New York: Viking, 1997), pp. 339–40.
23 Barry Smart, *Post-modernity* (London: Routledge, 1993), p. 35.
24 George Friedman and Meredith Friedman, *The Future of War: Power, Technology and American World Dominance in the Twenty-first Century* (New York: Crown Publishers, 1996), p. 10.
25 Cited in Richard Schlacht, 'The Future of Human Nature', in Paul Gifford (ed.), *2000 Years and Beyond: Faith, Identity and the 'Common Era'* (London: Routledge, 2003), p. 75.
26 Ibid., p. 76.
27 Ibid.
28 Richard Rorty, *Contingency, Irony and Solidarity* (Cambridge: Cambridge University Press, 1989), p. 170.
29 Octavio Paz, *Convergences: Essays in Art and Literature* (London: Bloomsbury, 1987), p. 139.
30 Richard Tarnas, *The Passion of the Western Mind: Understanding the Ideas that have Shaped Our World View* (London: Pimlico, 1991), p. 405.

31 Luc Ferry, *Man Made God: The Meaning of Life* (Chicago: Chicago University Press, 1996), p. 97.
32 Ibid., p. 19.
33 David Shukman, *The Sorcerer's Challenge: Fears and Hopes of the Weapons of the Next Millennium* (London: Coronet, 1995), p. 204.
34 Michel Foucault, *Discipline and Punish: The Birth of the Prison* (London: Penguin, 1977).
35 *The Times*, 3 September 2003.
36 Slavoj Žižek, *Welcome to the Desert of the Real* (New York: Verso, 2002), pp. 10–11.
37 Brian Rappert, *Non-lethal Weapons as Legitimizing Forces? Technology, Politics and the Management of Conflict* (London: Frank Cass, 2003), p. 41.
38 Raymond Tallis, *Enemies of Hope: A Critique of Contemporary Pessimism* (London: St Martin's Press, 1997), p. 67.

5 The Death of Sacrifice

1 Cited in John Ellis, *The Sharp End: The Fighting Man in World War II* (London: Pimlico, 1993), p. 322.
2 Peter Conrad, *Modern Times, Modern Places: Life and Art in the Twentieth Century* (London: Thames and Hudson, 1998), p. 102.
3 *The Times*, 12 February 2003.
4 *Star 21: Strategic Technologies for the Army of the Twenty-first Century*, Technology Forecast Assessments (Washington, DC: National Academy Press, 1993), p. 148.
5 Stephen Shaker and Alan Wise, *War Without Men: Robots on the Future Battlefield* (London: Pergamon Brasseys, 1988), p. 8.
6 See Conrad, *Modern Times, Modern Places*, p. 415.
7 John Horgan, *The End of Science: Facing the Limits of Knowledge in the Twilight of the Scientific Age* (New York: Little, Brown, 1996), p. 250.
8 Cited in Adam Roberts, *Science Fiction* (London: Routledge, 2000), p. 158.
9 Elaine Scarry, *The Body in Pain: The Making and Unmaking of the World* (New York: Oxford University Press, 1985), p. 152.
10 Hannah Arendt, *The Human Condition* (Chicago: Chicago University Press, 1998), p. 120.
11 Ibid., p. 121.
12 Jürgen Habermas, *The Future of Human Nature* (Cambridge: Polity, 2003), p. 38.
13 Manuel de Landa, *War in the Age of the Intelligent Machine* (New York: Zone, 1991), p. 130. There are, of course, other types of robots which will rely

much more on the biotechnological revolution. Robotic networks, colonies or nests of intelligent micro-robots (the product of nanotechnology), robots that are not singly intelligent like self-programming firing systems but collectively intelligent like ants in an ant colony, will rely on the biological imperatives of our own life: multiple interactions, positive and negative feedback, and amplified fluctuations. All are essential to self-organizing systems like an ant colony or the human body. Multiple interactions (such as a conversation) are what constitute society; feedback processes (e.g. what food we should eat) are essential to survival; fluctuations are how an event in one sphere of life impacts on every other (to take a military example, for want of a nail the battle was lost). Nanotechnology is simply the ability to do things – to measure or make – on the scale of atoms or molecules. It is a realm defined as being between 0.1 and 100 nanometres (a nanometre being one billionth of a metre). Nanotechnology could enable the military to breed colonies or nests of micro-intelligent robots to clear minefields or gather intelligence in reconnaissance missions. Such robots will be governed by simple behavioural rules, modelled on 'swarming' or other embodied evolutionary techniques. See Peter J Bentley, *Digital Biology* (London: Review, 2001), pp. 122–3.

6 To Be Concluded?

1 J. G. Ballard, *A User's Guide to the Millennium* (London: Flamingo, 1997), p. 23.
2 Steven Lukes, *Liberals and Cannibals: The Implications of Diversity* (London: Verso, 2003), pp. 73–4.

Index

Index

Index

Index

845

Index